SKELETON AND THE TEETH.

THE PRINCIPAL FORMS OF THE SKELETON AND OF THE TEETH.

BY

PROFESSOR R. OWEN, F. R. S., &c.,

AUTHOR OF "ODONTOGRAPHY;" "LECTURES ON COMPARATIVE ANATOMY;" "ARCHETYPE OF THE SKELETON;" "ON THE NATURE OF LIMBS;" "HISTORY OF BRITISH FOSSIL MAMMALIA," ETC. ETC.

PHILADELPHIA:
BLANCHARD AND LEA.
1854.

PHILADELPHIA:
T. K. AND P. G. COLLINS, PRINTERS.

PREFACE.

THE following work has just appeared in London as a portion of a series entitled *Orr's Circle of the Sciences.* Believing that a treatise of so much value was worthy of an independent position and a permanent form, the publishers issue it separately. Written by the most distinguished osteologist of the age, as an introduction to his favorite science, it cannot fail to possess great interest and value to all students of Zoology, Comparative Anatomy, and Geology, of which departments of knowledge Osteology may now be regarded as the foundation. As indicative of the principles which have guided the author in his labors, the following paragraphs are extracted from the Preface to the *Circle of the Sciences.*

"In regard to the structure and conformation of that great division of the Animal Kingdom called 'the Vertebrate,' to which Man himself belongs, and which includes the animals that most resemble Man, it has been deemed sufficient for present purposes to restrict the Essay to the fundamental structures or framework of the body, with the appendages of a like enduring material called the

Teeth. The execution of this part of the 'CIRCLE' has been confided to that great philosophical anatomist who has so distinguished himself in working out the true principles of Osteology—principles which will doubtless soon be applied to the nomenclature and description of every branch of Anatomical Science. Avoiding the common practice of intrusting the special essays to literary compilers and abridgers, it has been part of the design of the work—hitherto with success—to engage, in the important task of teaching, those master-spirits who have in their day effected the greatest improvements, and made the most decided advances, in their respective departments of science. The result has been, as is especially shown in the *Essay on the Principal Forms of the Skeleton*, an original exposition of the principles of Anatomical Science, and of the most important results that have been attained by its latest cultivator; such exposition being succinct without any important omission, and as clear and comprehensible as is consistent with the inevitable use of technical terms.

"New and clearly defined ideas must be expressed by their appropriate signs. The explanation of the sign teaches the nature of the idea. Without learning and understanding the technical terms of a science, that science cannot be comprehended. The terms seem 'hard' only while the ideas which they represent are not understood. We listen with pleasure and surprise to the glib facility with which the working classes, admitted in homely attire at half price to the Zoological Gardens on Mondays, talk of the Elephant, the Rhinoceros, and the Hippopotamus. These derivations from the Greek are no harder to them than the Saxon monosyllabic names of the bear, the seal, or the lion; and yet the four syllabled and five-syllabled names above cited are longer than the average of the technical terms derived from the same learned and pliable language: for example, Alisphenoid is not really harder than rhinoceros, nor Neurapophysis than hippopotamus; and when the mind becomes as familiar with the things of which these are the verbal signs, they fall naturally and easily into the circulating medium for the currency of thought. To the intelligent reader of every class, who may be blessed with the healthy desire for the attainment of knowledge, let it then be said: Be not dismayed with the array of 'hard words' which seems to bar your path in its acquisition. Where such words are invented or adopted by the masters in science, be assured that your acquisition and retention of their meaning will be the safest 'first steps' in the science of your choice.

"Where plain and known words of Saxon or old English root could convey the meaning intended, the writers have sedulously striven to use them instead of terms of more exotic origin. But where the signification of a thing, or group of things, would have demanded a roundabout explanation, or periphrase, as the alternative for abandoning the single-worded and clearly defined technical term, they have not hesitated to use such term, appending, either in the same page or in the Glossarial Index, its derivation and meaning.

"In reference to the terms of Anatomy, a method has been adopted further to facilitate their reception and easy recognition by reference to the part itself so signified in the wood-cuts; the same or corresponding part bearing the same numerals in all the cuts; thus the scapula, or blade bone, is indicated by the No. 51 in the fishes' skeleton, Fig. 9, and in all the succeeding skeletons up to those of the Ape and Man, Fig. 46."

CONTENTS.

ON THE PRINCIPAL FORMS OF THE SKELETON.

ON THE PRINCIPAL FORMS AND STRUCTURES OF THE TEETH.

LIST OF ILLUSTRATIONS.

2

THE

PRINCIPAL FORMS OF THE SKELETON.

PRINCIPLES OF OSTEOLOGY.

THE original substance of animals consists of a fluid with granules and cells. In the course of development, tubular tracts are formed, some of which become filled with "neurine," or nervous matter; others with "myonine," or muscular matter; other portions are converted into glandular substance; and a great proportion of the rest of the primordial matter forms "cellular substance." This substance, in many animals, becomes hardened, in certain parts of the body, by earthy salts. When those salts consist chiefly of phosphate of lime, the tissues called "osteine," or bone, and "dentine," are constituted, between which the chief distinction lies in the mode of arrangement of the earthy particles, in relation to the maintenance of a more or less free circulation of the nutrient juices through such hardened or calcified tissues. In bone, certain canals are left, of a caliber sufficient for the passage of capillary bloodvessels through the tissue. Still more minute tubes, sometimes expanding into cell-like cavities, are established for the slower percolation of the colorless fluid of the blood, called "plasma," or "liquor sanguinis." In true or hard dentine provision is made, by fine tubes, for the passage of plasma through its substance; but the red particles of the blood are excluded.

True osteine and dentine are peculiar to the highest division or province of the Animal Kingdom, which province has been termed "Vertebrata," from the prevalent disposition of the osseous matter in successive groups of more or less confluent bones called "vertebræ." (From *verto*, I turn; these being the parts on which the body bends or rotates.)

Before entering upon the disposition of the bony matter, a few words may be premised as to the composition of that matter in the different classes of vertebrata. These classes are four: Fishes, Reptiles, Birds, and Mammals, which latter class includes the hair-clad beasts, commonly called quadrupeds, with the naked whales and human kind. Fishes have the smallest proportion, birds the largest proportion, of the earthy matter in their bones. The animal part in all is chiefly a gelatinous substance.

PROPORTIONS OF EARTHY OR INORGANIC, AND OF ANIMAL OR ORGANIC, MATTER IN THE BONES OF THE VERTEBRATE ANIMALS.

FISHES.

	Salmon.	Carp.	Cod.
Organic	60.62	40.40	34.30
Inorganic . . .	39.38	59.60	65.70
	100.00	100.00	100.00

REPTILES.

	Frog.	Snake.	Lizard.
Organic	35.50	31.04	46.67
Inorganic . . .	64.50	69.96	53.33
	100.00	100.00	100.00

MAMMALS.

	Porpoise.	Ox.	Lion.	Man.
Organic	35.90	31.00	27.70	31.03
Inorganic . . .	64.10	69.00	72.30	68.97
	100.00	100.00	100.00	100.00

BIRDS.

	Goose.	Turkey.	Hawk.
Organic	32.91	30.49	26.72
Inorganic . . .	67.09	69.51	73.28
	100.00	100.00	100.00

From the above table, it will be seen that the bones of the fresh-water fishes have more animal matter, and are, consequently, lighter than those of fishes from the denser element of sea-water; and that the marine mammal called porpoise differs little from the sea-fish in this respect. The batrachian frog has more animal matter in its bones than the ophidian or saurian reptiles, and thereby, as in other respects, more resembles the fish. Serpents almost equal birds in the great proportion of the osseous salts, and hence the density and ivory-like whiteness of their bones.

The chemical nature of the inorganic or hardening particles, and of the organic basis of bone, is exemplified in the subjoined table, including a species of each of the four classes of vertebrata:—

CHEMICAL COMPOSITION OF BONES.

	Hawk.	Man.	Tortoise.	Cod.
Phosphate of lime with trace of fluate of lime . .	64.39	59.63	52.66	57.29
Carbonate of lime . . .	7.03	7.33	12.53	4.90
Phosphate of magnesia . .	0.94	1.32	0.82	2.40
Sulphate, carbonate, and chlorate of soda	0.92	0.69	0.90	1.10
Glutin and chondrin . .	25.73	29.70	31.75	32.31
Oil	0.99	1.33	1.34	2.00
	100.00	100.00	100.00	100.00

Bony matter is very variously disposed in the bodies of vertebrate animals. The sturgeon, the crocodile, and the armadillo, are instances of its accumulation upon or near the surface of the body; and hence the ball-proof character of the skin of the largest of these mailed examples. The most constant position of bone is around the central masses of the nervous and vascular systems, with rays thence extending into the middle of the chief muscular masses, forming the bases of the limbs. Portions of bone are also developed to protect and otherwise subserve the organs of the senses, and in some species are found encasing mucus-ducts, and buried in the substance of certain viscera; as, *e. g.*, the heart in the bullock and some other large quadrupeds. Strong membranes, called "aponeurotic," and certain leaders or tendons, become bony in some animals; as, *e. g.*, the "tentorium" in the cat, the temporal fascia in the turtle, the leaders of the leg-muscles in the turkey, the nuchal ligament in the mole, Fig. 41, *u*, and certain tendons of the abdominal muscles of the kangaroo, which, so ossified, are called the "marsupial bones," Fig. 44.

For a clear and intelligible view of the osseous system in general, it has become requisite to make a primary classification of its parts according to their prevalent position, as in the cases above cited. The superficial or skin-bones constitute the system of the "dermoskeleton" (from the Greek *derma*, skin, and *skeleton*); the deep-seated bones, in relation to the nervous axis and locomotion, form the "neuroskeleton" (Gr. *neuron*, nerve, and *skeleton*); the bones connected with the sense-organs and viscera form the "splanchnoskeleton (Gr. *splagchnon*, viscus, or inward part, and *skeleton*); those developed in tendons, ligaments, and aponeuroses, the "scleroskeleton" (Gr. *scleros*, hard, and *skeleton*). These technical terms may seem harsh, and sound strange, to those commencing the study of the structure of animals, but the most complex product of creation cannot be comprehended without terms expressive of the results of the classification and generalization of the manifold phenomena it offers to the contemplative student.

In the arrangement of the parts of the dermo, splanchno, and scleroskeleton, no common pattern is recognizable. One can discern a definite end or purpose gained by the positions those terms indicate of certain bony plates, cases, or rods, and the special relation of such to the habits and well-being of the creatures manifesting them; but the diversity in the number, size, shape, and relative position of such dermal bones and visceral bones seems interminable.

The head of the sturgeon, Fig. 1, is defended by a case of superficial bony plates, *d* 3, *d* 7, *d* 11, &c.; and the body, by five longitudinal rows of similar plates—one extend-

Fig. 1.

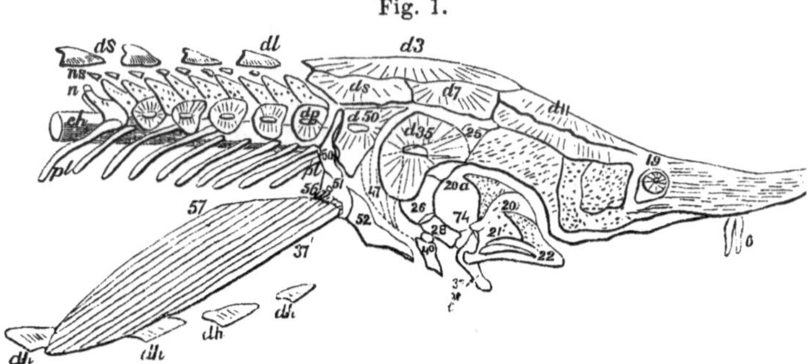

DERMO AND NEURO SKELETONS—STURGEON (*Acipenser Sturio*).

ing along the mid-line of the back, *ds*, *ds*, one along each side of the body, *dp*, *dp*, and two along the belly, *dh*, *dh*, between the fins called "pectoral," 57, and ventral. The observations of the ichthyologist, or of those concerned in the capture of the sturgeons for the sake of their air-bladder, of which the most valuable isinglass consists, show us how well the external defensive armor of these fishes is adapted to their mode of life. The sturgeons may be called the scavengers of the great rivers which they frequent. They habitually swim low, and grovel along the bottom, turning up the mud and sand with their pig-like snout, testing the disturbed matter with their feelers, 6, and feeding in shoals, on the decomposing animal and vegetable substances which are carried down with the debris of the continents drained by those rapid currents; thus, they are ever busied reconverting the substances, which otherwise would tend to corrupt the ocean, into their own living organized matter. These fishes are, therefore, duly weighted by a ballast of dense dermal, osseous plates—not scattered at random over their surface, but regularly arranged, as every seaman knows how ballast should be, in orderly series along the middle and sides of the body. The protection against the logs and stones hurried along their feeding-grounds, which the sturgeons derive from their plate armor, renders needless the ossification of the immediate case of the brain and spinal marrow, and, consequently, all the parts of the neuroskeleton, *ch*, *pl*, *n*, *ns*, remain in the flexible, elastic, gristly state; the weight of the dermoskeleton requiring that the other systems of the skeleton should be kept as light as might be compatible with its defensive and sustaining functions. This view of the final purpose of the dermal bony plates in the existing sturgeons affords some insight into the habits and conditions of existence of the similarly mailed extinct fishes which abounded in the seas of the secondary periods of the geological history of this planet. In most of these fishes, as in the sturgeons, the dermal bones are coated externally with a much harder material, resembling enamel, and such fishes have accordingly been termed "ganoid," from the Greek word "ganos," signifying brightness. The ganoid plates in those extinct fishes are usually more close-set, overlapping each other, and being fastened together like tiles, by a peg of one entering a socket in the next, and reciprocally. Only two genera of fishes are now known to exhibit this beautiful arrangement of the dermal bones, viz. the *polypterus* of the Nile, and the *lepidosteus* of the Ohio, and other great rivers of North America.

In the armadillo, the dermal bones, Fig. 2, *ob*, are small,

Fig. 2.

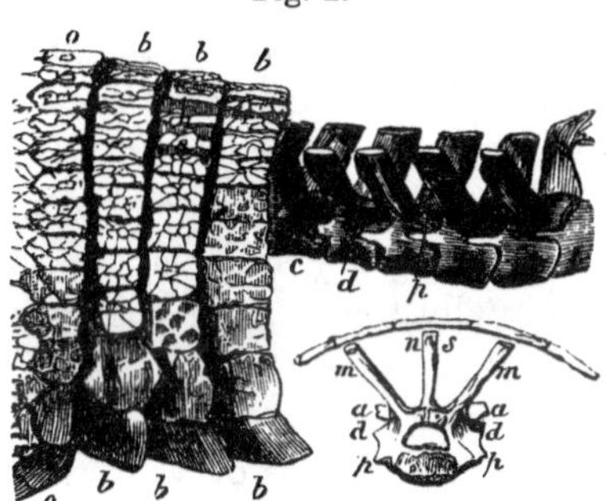

PORTIONS OF DERMO AND NEURO SKELETONS—ARMADILLO (*Dasypus tricinctus*).

polygonal, usually five or six-sided, smooth on their inner surface, which rests on the soft subcutaneous layer of cellular tissue, variously sculptured on the outer and exposed side, but with a pattern constant in, and characteristic of, each species. They are united together at their thick margins by rough or "sutural" surfaces, and resemble a tessellated pavement. The trunk is protected by a large buckler of this bony armor; the head is defended by a casque of the same; and the tail is encased in a sheath of similar interlocked ossicles. To allow of the requisite movements of the trunk in the small existing armadillos, which, when attacked, roll themselves into a ball, from three to nine transverse rows of the dermal bones, *b b*, are interposed, having a yielding elastic junction with each other, and with the anterior, *o o*, and posterior fixed, and larger, parts of the trunk-armor; and by this modification the head and limbs can be withdrawn beneath the armor, when its parts are pulled together by the strong cutaneous muscles into a hemispheric form. In South America, to which continent the armadillos are peculiar, remains of gigantic quadrupeds, similarly defended, have been discovered in the more recent tertiary deposits; but in these colossal armadillos (*Glyptodon*) the trunk-armor was in one immovable piece, covering the back and sides, and was not divided by bands. Besides the defence which such a modification of the integuments would afford against the attacks of predatory animals, the armadillos and glyptodens habitually frequenting the great forests of South America may have been protected by the same hard, arched covering from falling timber.

Such are some of the instances of the structure and uses of the dermoskeleton in the vertebrate province. The development of this system of the skeleton is not dependent on the grade of organization, for we find it in the highest and in the lowest classes; nor does a great amount of osseous matter in the skin necessarily involve a small amount or absence of the same matter in the deeper-seated skeleton; for all the parts of this system of bones, *a*, *c*, *d*, *m*, *ns*, are as well developed and as well ossified in the armadillos as in the quadrupeds which are covered by hair. The different states of the neuroskeleton, in the sturgeon and armadillo, are explicable only with reference to the different media and other conditions under which the two vertebrates were destined to exist.

In no species, and in no system of the skeleton, are bones a primary formation of the animal: they are the result of transmutations of pre-existing tissues, as substances composing animal bodies—*e. g.* nerve, muscle, membrane, &c.—are called. The inorganic salts, defined in the tabular view of the composition of bone, pre-exist in the blood, in the albumen of the egg of the oviparous vertebrates, and in the milk which nourishes the new-born mammal.

The primitive basis, or "blastema," of bone is a subtransparent glairy matter, containing a multitude of minute corpuscles. It progressively acquires increased firmness—sometimes assumes a membranous or ligamentous state, sometimes a gristly state, before its conversion into bone. Its assumption of the gristly state is attended by the appearance in it of numerous minute nucleated cells. As the gristle or "cartilage" hardens, these cells increase in number and size, and are aggregated in rows at the part where ossification is about to begin. These rows, in the cartilaginous basis of long bones, are vertical to its ends—in that of flat bones they are vertical to the peripheral edge. The nucleated cells are the instruments by which the earthy particles are arranged in order; and in bone, as in tooth, there may be discerned, in this predetermined arrangement, the same relation to the acquisition of strength and power of resistance, with the greatest economy of the building material, as in the disposition of the beams and columns of a work of human architecture.

Osteine, so formed, is arranged in thin plates, concentrically around the vascular canals, around the entire circumference of the long bones, and in interrupted plates, connecting together the walls of the vascular canals, so as often to give rise to a reticular disposition of the bony substance.

In fishes, the bones continue to grow throughout life; and their periphery, whether in the flat bones of the head which overlap each other, or the thicker bones that interlock, is cartilaginous or membranous, and the seat of progressive ossification. The long bones of most reptiles retain a layer of ossifying cartilage beneath the terminal articular cartilage; and growth continues at their extremities while life endures. Some of the long bones in frogs, birds, and most of those in mammals, have their ends distinct from the body or shaft of the growing bone, these separately ossified ends being termed "epiphyses." The seat of the active growth of the shaft is in a cartilaginous crust at the ends supporting the epiphyses; when these coalesce with the shaft, growth in the direction of the bones' axis comes to an end; but there is a slower growth going on over the entire periphery of the bone, which is covered by a membrane called the "periosteum." In this membrane, the vascular system of a bone, except the vessel supplying the marrow-cavity, undergoes the amount of subdivision which reduces its capillaries to dimensions suited for penetrating the pores leading to the vascular canals.

Thus bone is a living and a vascular part, growing by internal molecular addition and change, and having the power of repairing fracture or other injury. The shells and crusts of molluscous and crustaceous animals are unvascular; they grow by the addition of layers to their circumference, may be cast off when too small for the growing body, and be reproduced of a more conformable size. When fractured, the broken parts may be cemented together by newly superadded shell-substance from without; but are not unitable by the action of the fractured surfaces from within.

Extension of parts, however, is not the sole process which takes place in the growth of bone; to adapt a bone to its destined office, changes are wrought in it by the removal of parts previously formed. In fishes, indeed, we observe a simple unmodified increase. To whatever extent the bone is ossified, that part remains, and, consequently, most of the bones of fishes are solid or spongy in their interior, except where the ossification has been restricted to the surface of the primary gristly mould. The bones of the heavy and sluggish turtles and sloths, of the seals, and of the whale-tribe, are solid. But in the active land quadrupeds, the shaft of the long bones of the limbs is hollow, the first-formed osseous substance being absorbed as new bone is being deposited from without. The strength and lightness of the limb-bones are thus increased after the well-known principle of the hollow column, which Galileo, by means of a straw picked up from his prison floor, exemplified, in refutation of a charge of Atheism brought against him by the Inquisition. The bones of birds, especially those of powerful flight, are remarkable for their lightness. The osseous tissue itself is, indeed, more compact than in other animals; but its quantity in any given bone is much less, the most admirable economy being traceable throughout the skeleton of birds, in the advantageous arrangement of the weighty material. Thus, in the long bones, the cavities analogous to those called "medullary" in beasts, are more capacious, and their walls are much thinner. A large aperture called the "pneumatic foramen," near one end of the bone, communicates with its interior, and an air-cell, or prolongation of the lung, is continued into and lines the cavity of the bone, which is thus filled with rarefied air instead of marrow. The extremities of such air-bones present a light open network, slender columns shooting across in different directions from wall to wall, and these little columns are likewise hollow.

The enormous beak of the hornbill, which seems at first sight to constitute so grave an impediment to flight, forms one enormous air-cell, with very thin bony walls; and in this bird, in the swifts, and the humming-birds, every bone of the skeleton, down to the last joints of the toes, is permeated by hot air. The opposite extreme to the above members of the feathered class is met with in the terrestrial apteryx (wingless bird of New Zealand), and in the aquatic penguin; in both of which, not any bone of the skeleton receives air. Intermediate gradations in the extent to which the skeleton is permeated by air occur in different birds, and in relative proportion to their different kinds and power of flight.

In the mammalian class, the air-cells of bone are confined to the head, and are filled from the cavities of the nose or ear, not from the lungs. Such cells are called "frontal sinuses," "antrum," "sphenoidal" and "ethmoidal sinuses" in man. The frontal sinuses extend backwards over the top of the skull in the ruminant and some other quadrupeds, and penetrate the cores of the horns in oxen, sheep, and a few antelopes. The most remarkable development of air-cells in the mammalian class is presented by the elephant; the intellectual physiognomy of this huge quadruped being caused, as in the owl, not by the actual capacity of the brain-case, but by the enormous extent of the pneumatic cellular structure between the outer and inner plates of the skull-walls.

In all these varied modifications of the osseous tissue, the cavities therein, whether mere cancelli, or small medullary cavities as in the crocodile, or large medullary cavities as in the ox, or pneumatic cavities and sinuses as in the owl, are the result of secondary changes by absorption, and not of the primitive constitution of the bones. These are solid at their commencement in all classes, and the vacuities are established by the removal of osseous matter previously formed, whilst increase proceeds by fresh bone being added to the exterior surface. The thinnest-walled and widest air-bone of the bird of flight was first solid, next a marrow-bone, and finally became the case of an air-cell. The solid bones of the penguin, and the medullary femur of apteryx, exemplify arrested stages of that course of development through which the pneumatic wing-bone of the soaring eagle had previously passed.

But these mechanical modifications do not exhaust all the changes through which the parts of a skeleton, ultimately becoming bone, have passed; they have been previously of a fibrous or of a cartilaginous tissue, or both. Entire skeletons, and parts of skeletons, of vertebrate animals exhibit arrests of these early stages of development; and this quite irrespective of the grade of the entire animal in the zoological scale. The capsule of the eyeball, for example, in man, is a fibrous membrane; in the turtle, it is gristle; in the tunny, and most other fishes, it is bone. The skeletal framework of the little lancelet-fish (*Branchiostoma*) does not go beyond the fibrous stage of tissue change.[1] In the sturgeon, skate, and shark, it stops at the gristly stage, and hence these fishes are called "cartilaginous." In most fishes, and all air-breathing vertebrates, it proceeds to the bony stage, with the subsequent modifications and developments above recited.

The main part of the skeleton—what may be termed the skeleton proper—consists of the neuro-skeleton; and it is in the construction of this system that the most interesting and beautiful evidences of unity of plan, as well as of adaptation to end, have been discerned. The parts of the neuroskeleton are arranged in a series of segments, following and articulating with each other, in the direction of the axis of the body, from before backwards in brutes, from above downwards in man.

Each complete segment, called "vertebra," consists of a series of osseous pieces, arranged according to one and the same plan (Fig. 3), viz: so as to form a bony hoop, or arch, above a central piece, for the protection of a segment of the nervous axis, and a bony hoop, or arch, beneath the central piece, for the protection of a segment of the vascular system. The upper hoop is called the "neural arch," N. (Gr. *neuron*, nerve); the lower one, the "hæmal arch," H (Gr. *haima*, blood); their common centre is termed the "centrum," *c* (Gr. *kentron*, centre). The neural arch is formed by a pair of bones, called "neurapophyses," *n n* (Gr. for nerve and apophysis, a projecting part or process); and by a bone, sometimes cleft or bifid, called the "neural spine," *ns;* it also sometimes includes a pair of bones, called "diapophyses," *d d* (Gr. *dia*, across, or transverse, and apophysis). The hæmal arch is formed

[1] The doctrine or study of this kind of development—the development of substance and texture, as contradistinguished from that of size and shape—is now termed "Histology," from the Greek *histos*, net or tissue, and *logos*, a doctrine or discourse.

by a pair of bones called "pleurapophyses," *pl* (Gr. *pleuron*, rib, and apophysis); by a second pair, called "hæmapophyses," *h* (Gr. for blood, and apophysis; and by a bone sometimes bifid, called the "hæmal spine," *hs*. It also sometimes includes parts, or bones, called "parapophyses" (Gr. *para*, transverse, and apophysis). Bones, moreover, are developed, which diverge as rays, from one or more parts of a vertebra.

Fig. 3.

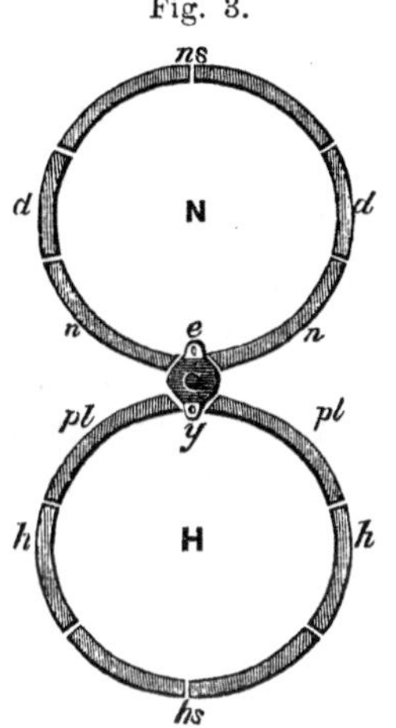

TYPICAL VERTEBRA.—(IDEAL.)

The parts of a vertebra which are developed from independent centres of ossification are called "autogenous;" those parts that grow out from previously ossified parts are called "exogenous;" the autogenous parts of a vertebra are its elements," the exogenous parts its "processes." No part, however, is absolutely autogenous throughout the vertebrate series, and some that are exogenous in most are autogenous in a few instances. The line cannot be strictly drawn; and, in classifying the parts of a vertebra, as of other parts of animals, or of entire animals, the systematist must be guided by general rules, to which there will ever be some exceptions.

The elements, or autogenous parts, of a vertebra are the centrum, *c*, the neurapophyses, *n*, the neural spine, *ns*, the pleurapophyses, *pl*, the hæmapophyses, *h*, and the hæmal spine, *hs*. The exogenous parts are the diapophysis (Fig. 5), *d*, the parapophysis (*ib.*) *p*, the zygapophysis (Fig. 6), *z* (Gr. *zugos*, junction, and apophysis), the anapophysis (Fig. 2), *a* (Gr. *ana*, backwards, and apophysis), the metapophysis (*ib.*), *m* (Gr. *meta*, between, and apophysis), the hypapophysis (Fig. 5), *y* (Gr. *hypo*, below, and apophysis), and the epapophysis (Fig. 4), *e* (Gr. *epi*, above, and apophysis). Of the autogenous parts, the neural spine is most commonly exogenous; of the exogenous parts, the parapophyses, diapophyses, and hypapophyses are sometimes autogenous.

Vertebræ are subject to many and great modifications —*e. g.*, as to the number of the elements retained in their composition, as to the form and proportion of the elements, and even as to the relative position of the elements; but the latter modification is never carried to such a degree as to obscure the general pattern or type of the segment.

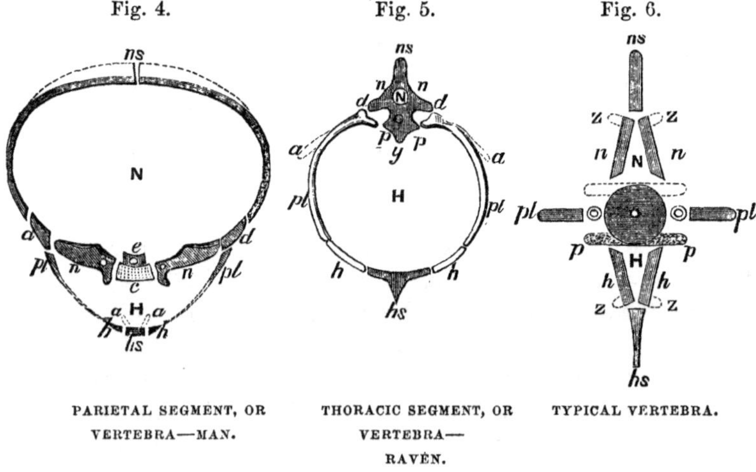

Fig. 4. PARIETAL SEGMENT, OR VERTEBRA—MAN.

Fig. 5. THORACIC SEGMENT, OR VERTEBRA—RAVEN.

Fig. 6. TYPICAL VERTEBRA.

Sometimes, as in the example (Fig. 4) of the third segment of the human skeleton, the neural arch, N, is much expanded, the hæmal one, H, is contracted; and, in the expanded neural arch, the autogenous diapophyses, *d d*, are wedged between the neurapophyses, *n*, and the enormously expanded neural spine, *ns*. More commonly, as in the example from the raven's thorax (Fig. 5), the hæmal arch, H, is much expanded, the neural one, N, contracted; and in the expanded hæmal arch, the parapophysis, *p*, here exogenous, is wedged between the centrum, C, and the pleurapophysis, *pl*. Sometimes, again, as is exemplified in the tail of the crocodile and of many other animals, both neural and hæmal arches are alike contracted; the pleurapophyses, *pl*, being excluded from the latter, and standing out as continuations of the confluent diapophyses, *d*, and

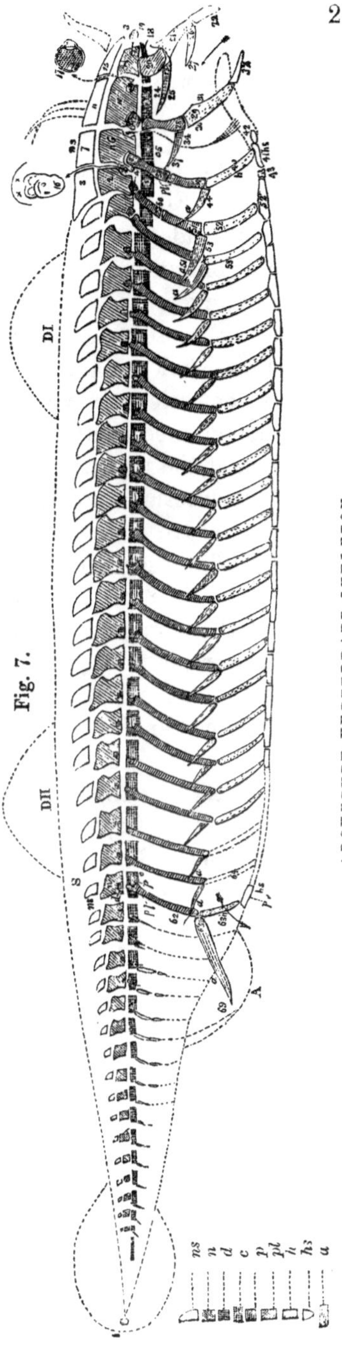

Fig. 7.

ARCHETYPE VERTEBRATE SKELETON.

parapophyses, *p*. Such vertebræ deviate but little from the ideal type of the vertebra, under its less developed condition, as in Fig. 6. The segments are commonly simplified and made smaller as they approach the end of the vertebral column or axis, one element or process after another is removed, until the vertebra is reduced to its centrum, as in the subjoined diagram (Fig. 7) of the archetype vertebrate skeleton. In this scheme, which gives a side view of the series of segments or vertebræ, the nature of the principal modifications to which they are subject are indicated, at the two extremes of the series.

As the four anterior divisions of the great trunk of the nervous system are called, collectively, "brain," so the four corresponding segments of the osseous system are called "skull." The head, therefore, is not otherwise a repetition of the trunk, than in so far as each segment of the skull is a repetition or "homotype" of every other segment of the body; each being subject to modifications which may give it an individual character, without obliterating its typical features. So neither are the "arms" and "legs" repeated in the head in any other sense than as the cranial vertebræ may retain their "diverging appendages," 25, 37, 44, 53, *a*. The fore-limbs are actually such appendages, 53, of the occipital vertebra, 1, 3, 2, 51, 52, 59, which appendages undergo modifications closely analogous to those of the appendages of the pelvic segment, or "hind limbs," 65. And inasmuch as in one class the pelvic appendages, with their supporting hæmal arch, 63, *hs*, are detached from the rest of their segment, and subject to changes of position (Fig. 9), 63, 69; so also in other classes the appendages of the occipital segment are liable to be detached, with their sustaining hæmal arch, and to be transported to various distances from their proper centrum and neural arch, as in Fig. 21, Nos. 51, 53, 57.

The four anterior neurapophyses, 14, 10, 6, 2, give issue to the nerves, the terminal modifications of which constitute the organs of special sense.

The first or foremost of these is the organ of smell, 19, always situated immediately in advance of its proper segment, which becomes variously and extensively modified to inclose and protect it.

The second is the organ of sight, 17, lodged in a cavity or "orbit" between its own and the nasal segment, but here indicated above that interspace.

The third is the organ of taste, the nerve of which perforates the neurapophysis, 6, of its proper segment, called "parietal vertebra," or passes by a notch between this and the neurapophysis, 10, of the frontal vertebra, to expand in the organ, which is always lodged below, in the cavity called "mouth," and is supported by the hæmal spine, 41, *hs*, of its own vertebra.

The fourth is the organ of hearing, 16, indicated above the interspace between the neurapophysis of its own (occipital) and that of the antecedent (parietal) vertebra, in which it is always lodged; the surrounding vertebral elements being modified to form the cavity for its reception, which is called "otocrane." The jaws are the modified hæmal arches of the first two segments.

The mouth opens at the interspace between these hæmal arches; the position of the vent varies (in fishes), but always opens behind the pelvic arch, S, 62, 63, *p*, when this is ossified.

Outlines of the chief developments of the dermoskeleton in different vertebrates, which are usually more or less ossified, are added to the neuroskeletal archetype; as, *e. g.* the median horn supported by the nasal spine, 15, in the rhinoceros; the pair of lateral horns developed from the frontal spine, 11, in most ruminants; the median folds, DI, DII, above the neural spines, one or more in number, constituting the "dorsal" fin or fins in fishes and cetaceans, and the dorsal hump or humps in the buffaloes and camels; similar folds are sometimes developed at the end of the tail, forming a "caudal" fin, C, and beneath the hæmal spines, constituting the "anal" fin or fins, A, of fishes.

The different elements of the primary segments are distinguished by peculiar markings:—

The neurapophyses by diagonal lines, thus— ////

The diapophyses by vertical lines— ||||||

The parapophyses by horizontal lines— ≡

The *centrum* by decussating horizontal and vertical lines— #

The pleurapophyses by diagonal lines— \\\\

The appendages by dots— : : : :

The neural spines and hæmal spines are left blank.

In certain segments, the elements are also specified by the initials of their names:—

ns is the neural spine.
n is the neurapophysis.
pl is the pleurapophysis.
c is the centrum.
h is the hæmapophysis, also indicated by the Nos. 21, 29, 44, 52, 58, 63, 64.
hs is the hæmal spine.
a is the appendage.

The centrum is the most constant vertebral element as to its existence, but not as to its ossification. There are some living fishes—and formerly there were many, now extinct—in which, whilst the peripheral elements of the vertebra become ossified, the central one remains unossified; and here a few words are requisite as to the development of vertebræ.

The central basis of the neuro-skeleton is laid down in the embryo of every vertebrate animal, as a more or less cylindrical fibrous sheath, filled with simple cells containing jelly. This fibro-cellulo-gelatinous column is called "notochord," Fig. 1, *ch* (Gr. *notos*, black; *chorda*, cord; in Latin, "chorda dorsalis"). The centrums, or "bodies of the vertebræ," as anthropotomists call them, are developed in and from the notochord. The bases of the other elements of the vertebra are laid down in fibrous bands, diverging from the notochord, and giving the first indication of the segmental character of the skeleton. At this stage, the skeleton of the little fish called "lancelet" (*Amphioxus lanceolatus*) is arrested. These fibrous bands are next converted into cartilage, and the cartilage is in definite pieces in each segment, recognizable as "neurapophyses" (Fig. 1), *n*; "pleurapophyses" (*ib.*), *pl*; "neural spine" (*ib*), *ns*—the centrums still remaining in their primitive state as the undivided notochord (*ib*), *ch*. At this stage, the skeleton of the sturgeon is arrested. The peripheral elements may be converted into bone, the central ones remaining as notochord, as in the protopterus, the lepidosiren, and many fossil fishes. But, more commonly, the next stage is the subdivision of the notochord into a series of separate centrums, corresponding with the pairs of neurapophyses and pleurapophyses—ossification of all the parts being more or less imperfect, as in the sharks and rays, which have thence been called "cartilaginous fishes." When the parts of the vertebræ have become more completely ossified, as in the fishes called "osseous," ossification is rarely so advanced as in the higher vertebrata. In most of these fishes, *e. g.*, a deep cavity is left at each end of the centrum (Fig. 8), *cc*, which cavity continues to be occupied by the liquefied gelatinous remains of the primitive notochord; and the characteristic of such element in a fish's skeleton is, that it is "biconcave." Of the minor amount of the earthy matter in the ossified parts of the skeleton of fishes, mention has been already made; and the consequent greater flexibility and elasticity of such bones may be readily tested by whoever will bend one of the long spines in the skeleton of a cod or turbot, and contrast its flexibility with that of the similarly-shaped long and slender bone (*pubis*, or *fibula, e. g.*) which he may find in the Christmas turkey that follows in the feast.

Fig. 8.

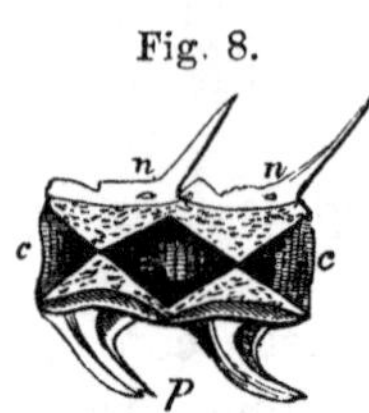

SECTION OF VERTEBRÆ —FISH.

Two or more contiguous vertebræ are frequently subjected to the same kind of modification, either by way of excess or defect, and such groups of modified segments have received special names; such, for example, as "skull" (*cranium*), "neck" (*cervix*), "chest" (*thorax*), "pelvis," and "tail" (*cauda*); and these terms are reciprocally applied, when modified as adjectives, to the individual vertebræ so grouped together, and which are called "cranial vertebræ," "cervical vertebræ," "dorsal" or "thoracic vertebræ," "sacral" or "pelvic vertebræ," and "caudal vertebræ."

SKELETON OF THE FISH.

In all fishes, the extent of ossification is less than in the higher vertebrate classes. Only in the skull do we find all the elements of the typical segment represented by bone. In the trunk, *e. g.*, the hæmapophyses and hæmal spines never advance beyond the fibrous stage of tissue development.

Four segments enter into the composition of the skull of fishes, answering to the first four in the archetype (Fig. 7), and they combine to constitute the bony framework of a head, larger in proportion to the trunk than in any other class of animals. The skull (Fig. 9), 3, 52, *br*, forms a cone, whose base is vertical, directed backwards, and joined to the trunk without an intervening neck, and whose sides are commonly three in number, one superior, and two lateral and inferior. The cone is shorter or longer, more or less compressed or squeezed from side to side, more or less depressed or flattened from above downwards, with a sharper or blunter apex, in different species of fishes. The base of the skull is perforated by the hole, called "foramen magnum," for the exit of the spinal marrow; the apex is more or less widely and deeply cleft transversely by the aperture of the mouth; the eye-sockets or "orbits," *or*, are lateral, large, and usually with a free and wide intercommunication in the skeleton; the two vertical fissures behind are called "gill-slits," or branchial or opercular apertures, and there is a mechanism, like a door, 34, 35, 36, for opening and closing them. The mouth receives not only the food, but also the streams of water for respiration (indicated by the arrow *br*), which escape by the gill-slits. The head contains not only the brain and organs of sense, but likewise the heart and breathing organs. The inferior or "hæmal" arches are greatly developed accordingly, and their diverging appendages support membranes that can act upon the surrounding fluid, and are more or less employed in locomotion; one pair of these appendages, P, 57, answers, in fact, to the fore-limbs in higher animals, and their sustaining arch, 51, 52, in many fishes, also supports the homologues of the hind-limbs, V, 69. Thus brain and sense-organs, jaws and tongue, heart and gills, arms and legs, may all belong to the head; and the disproportionate size of the skull, and its firm attachment to the trunk, required by these functions, are precisely the conditions most favorable for facilitating the course of the fish through its native element.

It may well be conceived, then, that more bones enter into the formation of the skull in fishes than in any other animals; and the composition of this skull has been rightly deemed the most difficult problem in comparative anatomy. "It is truly remarkable," writes the gifted Oken, to whom we owe the first clue to its solution, "what it costs to solve any one problem in philosophical anatomy. Without knowing the *what*, the *how*, and the *why*, one may stand, not for hours or days, but weeks, before a fish's skull, and our contemplation will be little more than a vacant stare at its complex stalactitic form."

To show *what* the bones are that enter into the composition of the skull of the fish; *how*, or according to what law, they are there arranged; and *why*, or to what end, they are modified, so as to deviate from that law or archetype, will next be our aim. These points, rightly understood, yield the key to the composition of the skull in all vertebrata, and they cannot be omitted without detriment to the main end of the most elementary essay on the skeletons of animals. The comprehension of the description will be facilitated by reference to Figs. 7 and 9; and still more if the reader have at hand the skull of any large fish.

In the cod (*Gadus morrhua*)[1], *e. g.*, it may be observed, in the first place, that most of the bones are, more or less, like large scales; have what, in anatomy, is called the "squamous" character and mode of union, being flattened, thinned off at the edge, and overlapping one another; and one sees that, though the skull, as a whole, has less freedom of movement on the trunk, more of the component bones enjoy independent movements. Before we proceed to pull apart the bones, it may be well to remark that the principal cavities, formed by their coadaptation, are the "cranium," lodging the brain and the organs of hearing; the "orbital," Fig. 9, *or*, and the "nasal," *nl*, chambers; the buccal and branchial canals, *br*. Some of these cavities are not well defined. The exterior of the skull is traversed by five longitudinal crests, intercepting four channels which lodge the beginnings of the great muscles of the upper half of the trunk. The median crest is developed to an extreme height in some fishes, as, *e. g.*, the dolphin and light-horseman fish (*Ephippus*). The flatfishes (turbot, sole, &c.) are remarkable for the unsymmetrical character of the skull, in consequence of both eyes being placed on one side of the head.

In the analysis of the cod's skull it is best to begin at the back part; for the segments of the skeleton deviate most from the archetype as they recede in position towards the two extremes of the body. After a little practice, one succeeds in detaching the bones which form the

[1] The skull of this fish, conveniently prepared for this examination, may be had of Mr. Flower, No. 22 Lambeth Terrace, Lambeth Road.

Fig. 9.

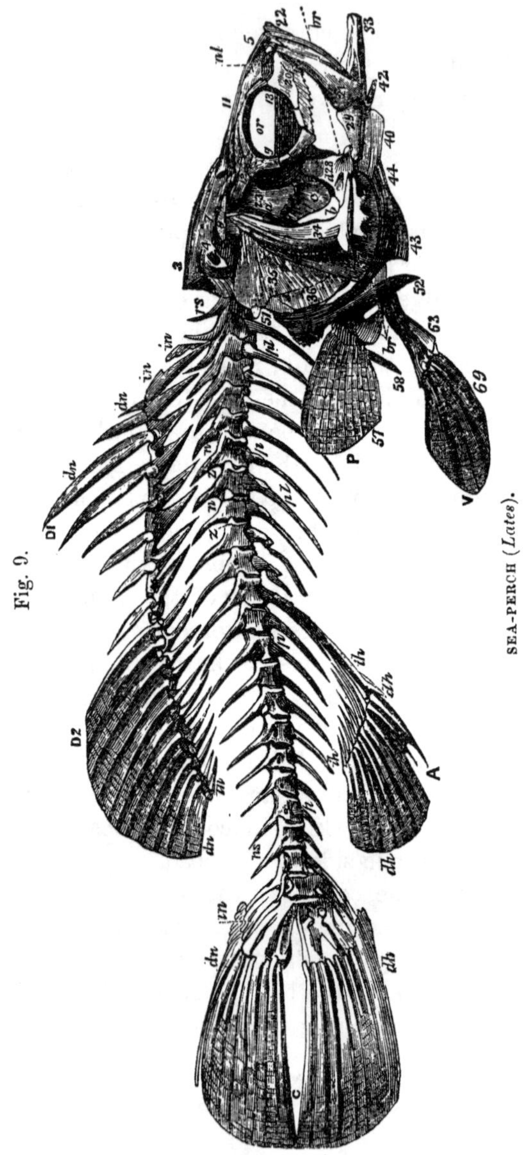

SEA-PERCH (*Lates*).

back part or base of the conical skull, and which immediately precede and join those of the trunk; we thus obtain a "segment" or "vertebra" of the skull. If we next proceed to separate a little the bones composing this segment, we find those that were most closely interlocked to be in number and arrangement as follows: Two single and symmetrical bones, and two pairs of unsymmetrical bones, forming a circle; or, if the lower symmetrical bone, which is the largest, be regarded as the base, the other five form an arch supported by it, of which the upper symmetrical bone is the keystone.[1] This answers to the "neural" arch of the typical vertebra: the base-bone is the "centrum," *c;* the pair of bones, which articulated with its upper surface and protected the hind division of the brain, form the "neurapophyses," *n;* the smaller pair of bones, projecting outwards, like transverse processes, are the "diapophyses," *d;* the symmetrical bone completing the arch, and terminating above in a long crest or spine, is the "neural spine," *ns.* It will be observed that the centrum is concave at that surface which articulates with the centrum of the first vertebra of the trunk; the opposite surface is also concave, but expanded and very irregular, in order to effect a much firmer union with the centrum of the next cranial segment in advance—great strength and fixity being required in this part of the skeleton, instead of the mobility and elasticity which is needed in the vertebral column of the trunk. It may be also observed that the "neurapophyses" are perforated, like most of those in the trunk, for the passage of nerves; that the diapophyses give attachment to the bones which form the great inferior or hæmal arch; and that the neural spine

[1] See my work "On the Archetype of the Skeleton," 8vo. 1848, p. 10, Fig. 1.

retains much of the shape of the parts so called in the trunk. Nevertheless, the elements of the neural arch of this hindmost segment of the skull have undergone so much development and modification of shape, that they have received special names, and have been enumerated as so many distinct and particular bones. The centrum, No. 1, is called "basioccipital;" the neurapophyses, No. 2, "exoccipitals;" the neural spine, No. 3, "superoccipital;" the diapophyses, No. 4, "paroccipitals." In the human skeleton all those parts are blended together into a mass, which is called the "occipital bone."

The entire segment, here disarticulated, in the cod-fish, is called the "occipital vertebra," and in it we have next to notice the widely expanded inferior or hæmal arch. This consists of three pairs of bones. The first pair are bifurcate, and have two points of attachment to the neural arch, the lower prong, answering to what is called the "head of the rib," abutting upon the neurapophysis; the upper prong, answering to the "tubercle of the rib," articulating to the diapophysis. The second pair of bones are long and slender, and represent the body of the rib. The first and second piece together answer to the element called "pleurapophysis; the third pair of bones are the "hæmapophyses;" these support diverging appendages consisting of many bones and rays. The special names of the above elements of the hæmal arch of the occipital vertebra are, from above downwards, "suprascapula," No. 50; "scapula," No. 51; "coracoid," No. 52. The inverted arch, so formed, encompasses, supports, and protects the heart or centre of the hæmal system; it is called the "scapular arch." There are animals—the gymnothorax and slow-worm, *e. g.*—in which this arch supports no appendage; there are fishes—the *protopterus*, *e. g.* Fig. 32—

in which it supports an appendage in the form of a single many-jointed ray, retaining the archetypal character, Fig. 7, No. 53. In other fishes, the number of rays progressively increase, until, in those called "rays" *par excellence*, they exceed a hundred in number, and are of great length, forming the chief and most conspicuous parts of the fish. The more common condition of the appendage in question is that exhibited in the species figured, Cut 9. So developed, it is called in ichthyology the "pectoral fin:" otherwise and variously modified in higher animals, the same part becomes a fore-leg, a wing, an arm, and hand. Some of the special names, originally applied to the parts of the scapular appendage in man, are retained and applied to like parts in the pectoral fin of the fish. Of the two flat bones connecting the fin with the coracoid, the upper one is the "ulna," No. 54; the lower one the "radius," No. 55; the row of short bones joined with these are the "carpals," No. 56; the longer and more slender many-jointed rays answer to the parts called "metacarpals" and "phalanges" in the human hand. In the salmon there is a bone answering to the arm-bone or humerus, which is articulated to the middle of the back part of the coracoid by a transversely elongated extremity. It is also expanded at the distal end, where it articulates by cartilage with the ulna and radius. The ulna is a semicircular plate of bone perforated in the centre, and, besides its articulation with the humerus, the radius, and the ulnar carpals and metacarpal ray, it also directly joins the broad coracoid. The radius, after expanding to unite with the humerus, the ulna, and the radial carpals, sends a long and broad process downwards and inwards, which is united by ligament with its fellow and with the lower termination of the coracoid. A basis of adequate extent and firmness is thus insured for the support of the pectoral fins. The carpal bones of these fins are four in number, progressively increasing in length from the ulnar to the radial side of the wrist. The meta-carpo-phalangial rays are thirteen in number; the uppermost or ulnar one being the strongest, and articulating directly with the ulna.

Proceeding to the next segment, in advance, in the cod-fish's skull, we find that the bone which articulated with the centrum of the occipital segment is continued forward beneath a great proportion of the skull. In quadrupeds, however, the corresponding part of the base of the skull is occupied by two bones; and if the single long bone in the fish be sawn across at the part where the natural suture exists in the beast, we have then little difficulty in disarticulating and bringing away with it a series of bones similar in number and arrangement to those of the occipital segment.

In the skeletons of most animals the centrums of two or more segments become, in certain parts of the body, confluent, or they may be connate; they form, in fact, one bone, like that, *e. g.*, which human anatomists call "sacrum." By the term "confluent" is meant the cohesion or blending together of two bones which were originally separate; by "connate," that the ossification of the common fibrous or cartilaginous bases of two bones proceeds from one point or centre, and so converts such bases into one bone: this is the case, *e. g.*, in the radius and ulna of the frog, and in its tibia and fibula. In both instances they are to the eye a single bone; but the mind, transcending the senses, recognizes such single bone as being essentially two. In like manner, it recognizes the "occipital bone" of man as essentially four bones; but these have become "confluent," and were not "connate." The centrums of the two middle segments of the fish's skull are connate, and the little violence above recommended is requisite to detach the penultimate segment of the skull. When detached, the bones of it are seen to be so arranged as to form a neural and a hæmal arch. In the neural arch the centrum, neurapophyses, diapophyses, and neural spine are distinct: moreover, the neural spine in the cod, and many other fishes, is bifid, or split at the median line.[1] The centrum is called "basisphenoid," No. 5; the neurapophysis, "alisphenoid," No. 6; the neural spine, "parietal," No. 7; and the diapophysis, "mastoid," No. 8. The alisphenoids protect the sides of the optic lobes, and the rest of the penultimate segment of the brain; the mastoids project outwards and backwards as strong transverse processes, and give attachment to the piers of the great inverted hæmal arch. Before noticing the structure of this, I may remark that, in the recent cod-fish, the case, partly gristly, partly bony, which contains the organ of hearing, is wedged in between the last and penultimate neural arches of the skull. The extent to which the ear-case is ossified varies in different fishes, but the bone is always developed in the outer wall of the case. In the cod-fish it is unusually large, and is called "petrosal," No. 16; it forms no part of the segmented neuroskeleton. In the organ which it contributes to inclose, there is a body as hard as shell, like half a split almond; it is the "otosteal," No. 16, or proper ear-bone.

The hæmal arch consists of a pleurapophysis and a hæmapophysis on each side, and a hæmal spine; but all these elements are subdivided; the pleurapophysis into two parts, the upper one called "epitympanic," 28, *a*

[1] "Archetype Vert. Skel.," p. 11, Fig. 2.

(common to this and the next arch in advance); the lower one "stylohyal," No. 38. The hæmapophysis is a broader, slightly arched bone; the upper division is called "epihyal," No. 39; the lower division, "ceratohyal," No. 40. The hæmal spine is subdivided into four stumpy bones, called collectively "basihyal," No. 41; and which, in most fishes, support a bone directed forwards, entering the substance of the tongue, called "glossohyal," No. 42; and another bone directed backwards, called "urohyal," No. 43.

The ceratohyal part of the hæmapophysis supports, in the cod, seven long and slender bent bones, called "branchiostegal rays," 44. The number of these rays differs greatly in different fishes; the *protopterus* has but one ray, the blenny has two rays, the carp three rays—a very common number is seven; but the *clops* has thirty branchiostegal rays. They are of great length in the angler-fish (*lophius*), in which they serve to support a membrane, developed to form a large receptacle on each side of the head of this singular fish. Into these receptacles the small fishes are transferred, which the angler attracts within reach of its mouth, by the movable rod, line, and bait attached to the top of its enormous head. In ordinary fishes, the branchiostegal rays support a membrane which helps to close the gill-slit, and by its movements contributes to the direction of the branchial currents. It is an appendage, or rudimental limb, answering to the pectoral fin diverging from the hæmal arch, in the adjoining occipital segment.

The penultimate segment of the skull above described is called the "parietal vertebra;" and the hæmal arch is called the "hyoidean arch," in reference to its supporting and subserving the movements of the tongue.

The next segment, or the second of the skull, counting

backward, can be detached from the foremost segment without dividing any bone. It is then seen to consist, like the third and fourth segments, of two arches and a common centre; but the constituent bones have been subject to more extreme modifications. The centrum, called "presphenoid," No. 9, is produced far forwards, slightly expanding; the neurapophyses, called "orbito-sphenoids," No. 10, are small semioval plates, protecting the sides of the cerebrum; the neural spine, or key-bone of the arch, called "frontal," No. 11, is enormously expanded, but in the cod and most fishes is single; the diapophyses, called "post-frontals," No. 12, project outwards from the hinder angles of the frontal, and give attachment to the piers of the inverted hæmal arch. The first bone of this arch is common in fishes to it and to that of the last-described vertebra, being the bone called "epitympanic," No. 28 (Fig. 9); this modification is called for by the necessity of consentaneous movements of the two inverted arches, in connection with the deglutition and course of the streams of water required for the branchial respiration. The hæmal arch of the present segment—enormously developed—is plainly divided primarily on each side into a pleurapophysis and hæmapophysis; for these elements are joined together by a movable articulation, whilst the bones into which they are subdivided are suturally interlocked together. The pleurapophysis is so subdivided into four pieces; the upper one, articulating with the post-frontal and mastoid—the diapophyses of the two middle segments of the skull—is called "epitympanic," No. 28, *a*; the hindmost of the two middle pieces is the "mesotympanic," No. 28, *b*; the foremost of the two middle pieces is the "pretympanic," No. 28, *c*; the lower piece is the hypotympanic, No. 28, *d*;

this presents a joint-surface, convex in one way, concave in the other, called a "ginglymoid condyle," for the hæmapophysis, or lower division of the arch. In most air-breathing vertebrates—the serpent, Cut 16, *e. g.*—the pleurapophysis resumes its normal simplicity, and is a single bone, 28, which is called the "tympanic;" in the eel-tribe it is in two pieces. The greater subdivision, in more actively breathing fishes, of the tympanic pedicle, gives it additional elasticity, and, by their overlapping, interlocking junction, greater resistance against fracture; and these qualities seem to have been required in consequence of the presence of a complex and largely-developed diverging appendage, which forms the framework of the principal flap or door, called "operculum," that opens and closes the branchial fissure on each side. The appendage in question consists of four bones; the one articulated to the tympanic pedicle is called "preopercular," No. 34; the other three are, counting downwards, the "opercular," No. 35; the "subopercular," No. 36; the "interopercular," No. 37. The hæmapophysis is subdivided into two, three, or more pieces, in different fishes, suturally interlocked together; the most common division is into two subequal parts, one presenting the concavo-convex joint to the pleurapophysis, and called "articular," No. 29; the other, bifurcated behind to receive the point of 29, and joining its fellow at the opposite end to complete the hæmal arch. It is very singularly modified by supporting, and having more or less firmly attached to it, a number of the hard bodies called "teeth," and hence it has been termed the "dentary," No. 33. In the cod there is a small separate bone, below the joint of the articular, forming an angle there, and called the "angular piece," No. 31.

In consequence of this extreme modification, in relation to the offices of seizing and acting upon the food, the pair of hæmapophyses of the present segment of the skull have received the name of "lower jaw," or "mandible" (*mandibula*). The entire segment is called the "frontal vertebra."

The first segment, forming the anterior extremity of the neuroskeleton, like most peripheral parts, is that which has undergone the most extreme modifications. The obvious arrangement, nevertheless, of its constituent bones, when viewed from behind, after its detachment from the second segment, affords one of the most conclusive proofs of the principle of adherence to common type which governs all the segments of the neuroskeleton, whatever offices they may be modified to fulfil. The neural arch plainly exists, but is now reduced to its essential elements—viz: the centrum, the neurapophyses, and the neural spine. The centrum is expanded anteriorly, where it usually supports some teeth on its under surface in fishes; it is called the "vomer," No. 13. The neurapophyses are notched (in the cod) or perforated (in the sword-fish), by the crura or prolongations of the brain, which expand into its anterior divisions, called "olfactory lobes;" the special name of such neurapophysis is "prefrontal," No. 14. The neural spine is usually single, sometimes cleft along the middle; it is the "nasal," No. 15.

The hæmal arch is drawn forwards, so that its apex, as well as its piers, are joined to the centrum (vomer) and usually also to the neural spine (nasal), closing up anteriorly the neural canal. The pleurapophyses are simple, short, sending backwards an expanded plate; they are called "palatines," No. 20. The hæmapophyses are simple, and their essential part, intervening between the pleurapophysis and hæmal spine, is short and thick; but they send a long process backwards. This element is called "maxillary," No. 21. The hæmal spine, cleft at the middle line, sends one process upwards, of varying length in different fishes, and a second downwards and backwards; and its under surface is beset with teeth in most fishes; it is called "premaxillary," No. 22. Each pleurapophysis supports a "diverging appendage," consisting commonly of two bones: the outer one, which fixes the present hæmal arch to the succeeding one, is called "pterygoid," No. 24; the inner one is the "entopterygoid," No. 23. The entire segment is called the "nasal vertebra." The hæmal arch and its appendage form what is termed the upper jaw (*maxilla*); the palatine and pterygoids forming the roof of the mouth, the maxillary and premaxillary the proper upper jaw. On reviewing the arrangement of the bones of the foregoing segments, one cannot but be struck by the strength of the arches which protect and encompass the brain, and by the beauty and efficiency of that arrangement which provides such an arch for each primary division of the brain; and a sentiment of admiration naturally arises on examining the firm interlocking of the extended sutural surfaces, and especially of those uniting the proper elements of the arch with the buttresses wedged in between the piers and keystone, and to which buttresses (diapophyses) the larger hæmal arches are suspended.

In addition to the parts of the neuroskeleton, the bones of the head include the ossified part of the ear-capsule, "petrosal," 16, already mentioned; an ossified part of the eye-capsule, commonly in two pieces, "sclerotals, No. 17; and an ossified part of the capsule of the organ of smell, "turbinal," No. 19. Another assemblage of splanchno-skeletal bones support the gills, and are in the form of slender bony hoops, called "branchial arches." They are articulated to and supported by the hyoidean arch. Amongst the bones of the muco-dermal system, may be noticed those that circumscribe the lower part of the orbit, of which the anterior is pretty constant in the vertebrate series, and is called "lachrymal," marked 20 in Cut 9. In fishes, they are called "suborbitals," and are occasionally present in great numbers, as *e. g.*, in the tunny. A similar series of bones sometimes overarches the temporal fossæ, and are called "supertemporals."

At the outset of the study of Osteology, it is essential to know well the numerous bones in the head of a fish, and to fix in the memory their arrangement and names. The latter, as we have seen, are of two kinds, as regards the bones of the neuroskeleton; the one kind is "general," indicative of the relation of the skull-bones to the typical segment, and which names they bear in common with the same elements in the segments of the trunk; the other kind is "special," and bestowed on account of the particular development and shape of such elements, as they are modified in the head for particular functions. I would advise any one earnestly desirous of comprehending this beautiful department of Comparative Anatomy, to obtain a prepared and partially disarticulated skull of a cod-fish from Mr. Flower,[1] in which every bone bears the initials of its "general" name, and the numerals indicative of its "special" name. A great proportion of the bones in the head of a fish exist in a very similar state of connection and arrangement in the heads of other vertebrata, up to and including man himself. No method could be less

[1] *Ante*, p. 37.

conducive to a true and philosophical comprehension of the vertebrate skeleton than the beginning its study in man—the most modified of all vertebrate forms, and that which recedes furthest from the common pattern. Through an inevitable ignorance of that pattern, the bones in anthropotomy are indicated only by special names more or less relating to the particular forms these bones happen to bear in man; such names, when applied to the tallying bones in lower animals, losing that significance, and becoming arbitrary signs. Owing to the frequent modification by confluence of the human bones, collections of them, so united, have received a single name, as *e. g.*, "occipital," "temporal," &c.; whilst their constituents, which are usually distinct vertebral elements, have received no names, or are defined as processes, *e. g.*, "condyloid process of the occipital bone," "styloid process of the temporal bone," "petrous portion of the temporal bone," &c. The classification, moreover, of the bones of the head in Human Anatomy, viz: into those of the cranium and those of the face, is artificial or special, and consequently defective. Many bones which essentially belong to the skull are wholly omitted in such classification.

In regard to the archetype of the vertebrate skeleton, fishes, which were the first forms of vertebrate life introduced into this planet, deviate the least therefrom; and according to the foregoing analysis of the bones of the head, it follows that such bones are primarily divisible into those of—

The Neuroskeleton;
The Splanchnoskeleton;
The Dermoskeleton.

The neuroskeletal bones are arranged in four segments, called

The Occipital segment;
The Parietal segment;
The Frontal segment;
The Nasal segment.

Each segment consists of a "neural" and a "hæmal" arch. The neural arches are—

N I. Epencephalic arch (bones Nos. 1, 2, 3, 4):
N II. Mesencephalic arch (5, 6, 7, 8);
N III. Prosencephalic arch (9, 10, 11, 12);
N IV. Rhinencephalic arch (13, 14, 15).

The hæmal arches are—

H I. Scapular arch (50–52);
H II. Hyoidean arch (38–43);
H III. Mandibular arch (28–32);
H IV. Maxillary arch (20–22).

The diverging appendages of the hæmal arches are—

1. The Pectoral (54–57);
2. The Branchiostegal (44);
3. The Opercular (34–37);
4. The Pterygoid (23–24).

The bones or parts of the splanchnoskeleton which are intercalated with or attached to the arches of the true vertebral segments, are—

The Petrosal (16) or ear-capsule, with the otolites, 16″;
The Sclerotal (17) or eye-capsule;
The Turbinal (19) or nose-capsule;
The Branchial arches;
The Teeth.

The bones of the dermoskeleton are—

The Supratemporals;
The Superorbitals;
The Suborbitals;
The Labials.

Such appears to be the natural classification of the parts which constitute the complex skull of osseous fishes.

The term "cranium" might well be applied to the four neural arches collectively, but would exclude some bones called "cranial," and include some called "facial," in Human Anatomy. In a side view of the naturally-connected bones of the head of a fish, such as is shown in the figure of the skeleton of the sea-perch, Cut 9, the upper part of the head is formed by the neural spines called superoccipital, 3, frontal, 11, and nasal, 15; produced at the hinder half into the median ridge. The right lateral ridge is formed by the parietal, 7, and paroccipital, 4; the external ridge by the post-frontal, 12, and the mastoid, 8. The anterior termination of the series of centrums may be partly seen through the widely-open orbits at 9 and 13, indicating the presphenoid and vomer respectively. The most conspicuous parts of the upper jaw are the premaxillary, 22, and the maxillary, 21, the latter being edentulous, as in most fishes; the salmon and trout are examples where No. 21 bears teeth. The shape and slight attachment of those bones relate to the necessity of a movable mouth that can be protruded and retracted, in a class of animals that derive no aid in the prehension of their food from their limbs, which are reduced to fins. The upper bent back part of the premaxillary is called its "nasal branch," and is of unusual length in fishes with protractile snouts, as, *e. g.*, the dories (*Zeus*), certain wrasses (*Coricus*), and especially the sly-bream (*Sparus insidiator* of Pallas). In this fish, the nasal branch of the premaxillary plays in a groove on the upper surface of the skull, and reaches as far back as the occiput, when the mouth is shut and retracted. The descending branch of the premaxillary is attached by a ligament to the maxillary, and, as this is similarly attached to the mandible, both are protruded, when the long nasal branch of the premaxillary is drawn forwards out of its epicranial groove. This action is aided by the hypotympanic, which is of great length, and has a movable articulation at both ends; the lower end joining the mandible is pulled forward, simultaneously with the protrusion of the premaxillary, and co-operates therewith in the sudden projection of the mouth, by which the sly-bream seizes, or shoots with a suddenly-propelled drop of water, the small agile aquatic insects that constitute its prey.

An opposite extreme of modification of the maxillary and premaxillary bones, where unusual fixity and strength are needed, is that presented by the "sword-fishes," in which the premaxillaries constitute, by an unusual prolongation and density of tissue, the sword-shaped weapon characteristic of the genera *Xiphias* and *Istiophorus.*

In Cut 9 the divisions 28 *a*, *c*, and *d*, of the tympanic pedicle, and the two chief divisions, 29 and 33, of the mandible, are shown, together with the four bones of the opercular appendage; the preopercular, 34, being serrated and spined, as in most perches.

Of the hyoidean arch may be seen the glossohyal, 42, the ceratohyal, 40, with its branchiostegal rays, 44, and the urohyal, 43. Of the scapular arch, the scapula, 51, and the coracoid, 52, this supports not only the bones of the "pectoral fin," P, viz: ulna, radius, with the small carpal bones intervening between them and the metacarpophalanges, 57, but also the lower elements of the pelvic arch, 63, and their diverging appendage, 69, called the "ventral fin," V.

In the segments of the trunk the hæmapophyses, save in the first vertebra, 58, and the pelvic vertebra, 63, are not ossified; but they are represented by aponeurotic fascia continued downwards from the ossified elements of

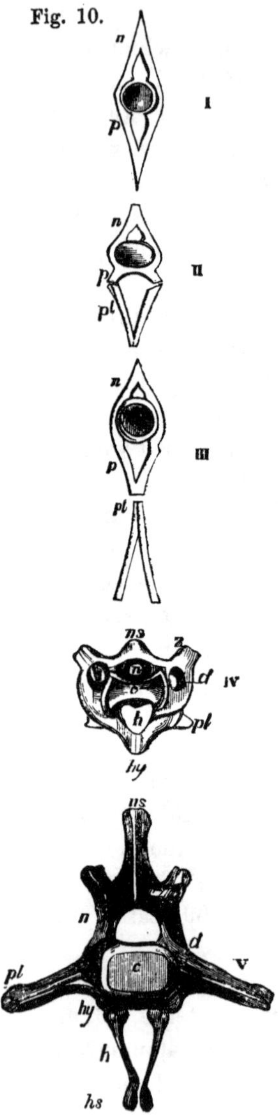

Fig. 10.

the segments; these elements consist of the centrum, the neurapophyses, and neural spines, the pleurapophyses, and the parapophyses. In most fishes the neural spines are connate with the neurapophyses, and these become confluent with the centrums in most of the segments; the neurapophyses are perforated directly by the spinal nerves in many fishes, as at *nn* (Fig. 8); they usually develop anterior zygapophyses, Z (Fig. 9). The centrums are biconcave in all fishes, save the lepidosteus, in which they are convex in front, and concave behind. The pleurapophyses, *pl* (Fig. 9), form what are called "false ribs," or free or "floating ribs," in Anthropotomy; they articulate with the centrums in the anterior trunk-vertebræ, and then with the parapophyses, *p*, which are usually confluent with the centrum. The parapophyses elongate, bend down, and unite together at or near to the end of the abdomen, and so form the contracted hæmal canal, for the caudal vessels, in the long and muscular tail of the fish. The trunk-vertebræ of a fish are divisible into those which have free pleurapophyses, called "abdominal vertebræ," and those without, and which terminate below by narrow hæmal arches and long spines, called "caudal vertebræ." These hæmal arches are formed by different parts in different fishes; commonly by the bent-down and terminally confluent parapophyses (Fig. 10), I, *p*, cod; sometimes, as in the tunny (*ib.*) III, by parapophyses, *p*, lengthened out by pleurapophyses, *pl*; sometimes, as in lepidosteus (*ib.*), II, by pleurapophyses, *pl*; but never, as in air-breathing vertebrates (*ib.*), V, by ossified hæmapophyses, *h*, *hs*. These elements, in the first vertebra of the trunk of a fish, are indeed ossified, and form the long and slender bone called "clavicle," 58 (Fig. 9), usually attached to the inner side of the scapular arch. The hæmapophyses of, probably, the last abdominal vertebra, called "ischia," No. 63, are detached from the rest of their segment, and are either loosely suspended in the flesh, beneath or near it, as in the fishes called "abdominal;" or they are advanced, much elongated, and attached to the scapular arch, as in the fishes called "thoracic" (Fig. 9); or they are more advanced, shortened, and similarly attached, as in the fishes called "jugular;" or they are wholly wanting, as in the fishes called "apodal." The fins called "ventral," V, supported by the pelvic hæmapophyses, indicate by their position the orders of fishes called "abdominal," "thoracic," and "jugular," by Linnæus.

The only proper fins in pairs are the "pectoral," P, answering to the fore-limbs of quadrupeds, and the "ventral," V, answering to the hind-limbs. The rest of the fins are single and median in position, and are due to folds of the skin, in which certain dermal bones are developed for their support. These bones are of two kinds; one, dagger-shaped, are plunged, so to speak, up to the hilt, in the flesh between the neural spines, and between the hæmal spines; those along the upper surface of the fish are called "interneural spines," *in*, Cut 9; those on the under surface are the "interhæmal spines," *ih*. The interneural spines support the "dermoneural spines, *dn*, forming the rays of the dorsal fin or fins, D1, D2, and the upper rays of the caudal fin. The interhæmal spines support the dermohæmal spines, *dh*, which form the rays of the anal fin, A, and the lower rays of the caudal fin, *dh*, C.

Both dermoneural and dermohæmal spines may present two structures; they may be simple, unjointed, firm, bony spines; or they may be flexible, jointed, and branched rays. Those fishes which have one or more of the hard spines at the beginning of the pectoral, ventral, dorsal, and anal fins are called "acanthopterygian," or spiny-finned fishes (Gr. *acanthos*, spine; *pterux*, fin); those in which the vertical fins are supported by soft spines are called "malacopterygian," or soft-finned fishes (Gr. *malakos*, soft; and *pterux*). Ichthyologists avail themselves of the number and kind of rays in the fins to characterize the species of fishes, and adopt an abbreviated formula and symbols to express these characters.

In regard to the sea-perch (Fig. 9), the fin-formula would be as follows:—

D7,1+12 : P12 : V1+5 : A3+8 : C18,

which signifies that D, the dorsal fin, has, in its first division, 7 rays, all spinous; in its second division, 1 spinous + (plus) 12 rays that are soft. P, the pectoral fin, has 12 rays, all soft. V, the ventral fin, has one spinous+5 soft rays. A, the anal fin, has 3 spinous+8 soft rays. C, the caudal fin, has 18 rays.

When the piscine modification of the vertebrate skeleton is contemplated in relation to the life and movements of a fish in its native element, every departure from the archetype is seen to be in direct relation to the habits and well-being of the species.

The large head has been compared to the embryonic disproportion of that part in higher vertebrates; but the head of a fish should be of the size and shape best fitted to overcome the resistance of water, and to facilitate rapid progression through that element; the head must, therefore, grow with the growth of the body. Accordingly, the large skull-bones always show the radiating bony filaments in their clear circumference, which is the seat of growth; and hence the number of overlapping squamous sutures which least oppose the progressive extension of the bones. The cranial cavity expands with the expansion of the skull, but the brain undergoes no corresponding increase; it lies at the bottom of its capacious chamber, which is principally occupied by a loose cellular tissue, situated, like the "arachnoid" membrane in man, between the brain-tunics, called "pia mater" and "dura mater," and having its cells filled by a light, oily fluid; thus the head is rendered specifically lighter than if growth only, and not the modelling absorption also, had gone on. The loose connection of the hæmal arches and their parts, including most of what are called "bones of the face," seems like the retention of a condition observable in the partially-developed skull of the embryos of higher animals; but this condition is subservient to the peculiar and extensive movements of the jaws, and of the bony supports of the breathing machinery. Not any of the limbs of fishes are prehensible; the mouth may be propelled by them to the food, but the act of taking it must be performed by the jaws; these can, accordingly, be not only opened and shut, but can be protruded and retracted. The division of the long tympanic pedicle into several partly-overlapping pieces adds to its strength, and by a slight elastic yielding diminishes the liability to fracture. The tongue, to judge by its structure, seems to serve little as an organ of taste, but the arch sustaining it has much to effect in the way of swallowing; for this action relates not merely to food; the mechanical part of breathing is a modified, habitual, and frequent act of deglutition. The hyoid arch is the chief support of the branchial arches and gills; and the branchiostegal membranes, stretched out upon the diverging rays of the hyoid arch, regulate the course and exit of the respiratory currents.

By the retraction of the hyoid arch the opercular doors are forced open, and the branchial cavity is widened, whilst all entry from behind is prevented by the branchiostegal flaps, which close the external gill-openings. The water, therefore, enters by the gaping mouth, and rushes through the sieve-like interspaces of the branchial arches into the branchial cavity; the mouth then shuts, the opercular doors close upon the branchial and hyoid arches, which again swing forwards; and the branchiostegal membranes being withdrawn, the currents rush out at the gill-openings. Thus the mechanical functions of the hæmal arches of the thorax of the higher air-breathing classes are transferred to the hæmal arches and appendages of the skull in fishes.

The persistent gills and gill-arches in fishes have been compared with the same parts which are transitory in frogs, and with some traces of branchial organization in the embryos of higher vertebrates; and fishes have been called, in the language of the transmutation-of-species hypothesis, "arrested gigantic tadpoles." It will be found, however, that so far from there having been any stoppage of development, the branchial arches have been adapted to the exigencies of the fish by advancing to a grade of structure which they never reach in the frog. This is shown by their firm ossification, and their numerous elastic joints; the sieve-like valves developed from the side next the mouth have been pre-arranged, with the utmost complexity and nicety of adjustment, to prevent the entry of any particles of food, or other irritating matters, into the interspaces of the tender, vascular, and sensitive gills. It is interesting, also, further to observe, that the last pair of these arches, which, when the embryo-fish is as yet edentulous, usually support gills, are reduced, when the supply of yolk-food is exhausted, and the jaws get their prehensile organs, to the capacity of the gullet, become thickened, in order to support teeth for tearing in pieces, mincing, or crushing the food, and are converted into an accessory pair of jaws, and this pair the most important of the two, as it would seem; for the carp-tribe—*e. g.*, tench, barbel, roach—which have no teeth on their proper jaws, have teeth on the pharyngeal jaws. In no other vertebrate animals, save the osseous fishes, is the mouth provided with maxillary instruments at both the fore and hind apertures; and in no other part of the piscine structure is the direct divergence from any conceivable progressive scale of ascending organisms, culminating in man, so plainly marked as in this.

The general form of the fish is admirably adapted to the element in which it lives and moves. The viscera are packed in a moderate compass, in a cavity brought forwards close to the head. The absence of any neck gives the advantage of a more extensive and resisting attachment of the head to the trunk, and a greater proportion of the trunk is left free for the allocation of the muscular masses which move the tail. In the "caudal" division of the vertebral column, the parapophyses cease to extend outwards; they bend downwards, unite and elongate in that direction, proportionally with the elongation of the spines above, whilst dermal and intercalated spines shoot forth from the middle line above and below, giving the vertically extended, compressed form to the hinder half of the body, by the alternating lateral strokes of which the fish is propelled forwards in the diagonal between the direction of those forces. The advantage of the biconcave form of vertebra, with intervening elastic capsules of gelatinous fluid, in producing a combination of the resilient with the muscular power, is as obvious as it is beautiful to contemplate.

The fixation and coalescence of any of the vertebra in this locomotive part of the fishes' body, analogous to the part called "sacrum" and "pelvis" in land quadrupeds, would be a great hindrance to the alternate and vigorous inflections of that part, by which mainly the fish swims. A "sacrum" is a consolidation of part of the vertebral axis of the body, for the transference of more or less of the weight of that body upon limbs organized for its support on dry land; such a modification would have been not merely useless, but a hinderance to a fish. The pectoral fins are the prototypes of the fore-limbs of the higher vertebrates. With their terminal segment, or "hand," alone projecting freely from the trunk, and swathed in a common sheath of skin, they present an interesting analogy to the embryonal buds of the answerable members in man. But what would have been the result if both arm and forearm had extended freely from the side of the fish, and dangled as a long many-jointed appendage in the water! This "higher development," as it is termed, in relation to the prehensile or cursorial limb of the denizen of dry land, would have been a defect in the structure of a creature destined to cleave the liquid element. In the fish, therefore, the fore-limb is left as short as was compatible with its required functions; the broad many-fingered hand alone projects, but can be applied prone and flat, by flexion of the wrist, to the side of the trunk; or it may be extended with its flat surfaces turned forwards and backwards, so as to check and arrest, more or less suddenly, the progress of the fish; its breadth can also be diminished by closing up or stretching out the digital rays. In the act of flexion, the pectoral fin slightly rotates, and gives an oblique stroke to the water. The requisite breadth of the modified hand is gained by the addition of ten, twenty, or it may be a hundred fingers over and above the number to which they are restricted in the forefoot or hand of the higher classes of vertebrata. The pike maintains a stationary position in a stream by vibrations of the pectoral fins; the nature of the bottom of the fish's habitat is ascertained by a tactile application of the same fins. In the hard-faced gurnards, certain rays of the pectorals are liberated from the web, and have a special endowment of nerves, in order to act as feelers. In the siluroid fishes, the pectorals wield a formidable weapon of offence. A tropical species of perch (*Anabas*) uses a smaller analogous pectoral spine for climbing up the mangrove stems in quest of insects.

Certain lophioid fishes, that live on sand-banks left dry at low water, are enabled to hop after the retreating tide by a special prolongation of the carpal joint of the pectoral fin, which projects in these "frog fishes," as they have been termed, like the limb of a land quadruped, and presents two distinct segments clear of the trunk.

The sharks, whose form of body and strength of tail enable them to swim near the surface of the ocean, are further adapted for this sphere of activity, and compensated, for the absence of an air-bladder, by the large proportional size of their pectoral fins, which take a greater share in their active and varied evolutions than in ordinary fishes; more especially in producing that half turn or roll of the body required to bring the mouth, which is on the under part of the head, in contact with their prey. The maximum of development of the many-fingered hands is attained in the rays, and in those fishes—*e. g.*, *Exocœtus* and *Dactylopterus*—called "flying fishes," in consequence of the pectorals being long enough, and their webs broad enough to sustain them in the air, in their long "flying leaps" out of the water.

With regard to the ventral fins—the rudiments of hind-limbs—these combine merely with the pectorals in raising the fish, and in preventing, as outriggers, the rolling of the body during progression. In the long-bodied and small-headed abdominal fishes the ventrals are situated near the vent, where they best subserve the office of accessory balancers; in the large-headed thoracic and jugular fishes they are transferred forwards, to aid the pectorals in supporting and raising the head. If the pectoral and ventral fins in one of these fishes be cut off, the head sinks to the bottom; if the right pectoral fin only be cut off, the fish leans to that side; if the ventral fin on the same side be cut away, then it loses its equilibrium entirely; if the dorsal and anal fins be cut off, the fish reels to the right and left; when the caudal fin is cut off, the fish loses the power of progressive motion; when the fish dies, and the fins cease to play, the belly turns upwards.

Paley thus sums up the actions of the fins of fishes: "The pectoral, and more particularly the ventral, fins serve to raise and depress the fish; when the fish desires to have a retrograde motion, a stroke forward with the pectoral fin effectually produces it; if the fish desire to turn either way, a single blow with the tail the opposite way sends it round at once; if the tail strike both ways, the motion produced by the double lash is progressive, and enables the fish to dart forwards with an astonishing velocity. The result is not only in some cases the most rapid, but in all cases the most gentle, pliant, easy animal motion with which we are acquainted." "In their mechanical use, the anal fin may be reckoned the keel; the ventral fins, the outriggers; the pectoral fins, the oars;" and we may now add "the caudal fin, the screw-propeller." And if there be such similitude between those parts of a boat and a fish, "observe," adds Paley, "that it is not the resemblance of imitation, but the likeness which arises from applying similar mechanical means to the same purposes."[1]

[1] "Nat. Theology," 8vo., 1805, p. 257.

PRINCIPAL FORMS OF THE SKELETON IN THE CLASS REPTILIA.

The transition from fishes to reptiles is easy, and the signs thereof very manifest in the skeleton. In the thorn-back and allied fishes the skull articulates with the trunk by two condyles, and the part answering to the basioccipital is a depressed plate. The *Batrachia*, or lowest order of reptiles—including the siren, proteus, frog, toad—have a similar double articulation of the skull with the trunk, the two condyles being developed from the two exoccipitals. Hæmapophyses are not present as bones in the abdominal part of the trunk of *Batrachia*, but they are so developed in the tail. This structure, with the detachment of the scapular arch from the occiput, and the absence of dermoneural and dermohæmal spines, serves to distinguish the most fish-like batrachian from the protopterus and lepidosiren, which are the most reptile-like of fishes.

In commencing the study of the skeletons of reptiles in the most fish-like of the class, we find a much less complex condition of the osseous framework of the body than in the bony fishes; this will be immediately manifest by a comparison of the skeleton of the menopome (which may be seen in the Museum, Royal College of Surgeons, No. 583), as an example of the perennibranchiate batrachia, with the skeleton of the trout (No. 45) or of the haddock (No. 176, in the same Museum).

The difference tends greatly to elucidate the true nature of the complexities of the fish's skeleton, since it chiefly consists in the simplification of that of the batrachian, by the non-development of the parts of the dermal skeleton which characterize that of the fish. The suborbital, superorbital, and supratemporal scale-bones are removed, together with the opercular bones, from the head; and the interneural and dermoneural spines, with the interhæmal and dermohæmal spines, are removed from the trunk. The endoskeleton is also reduced to a very simple condition; the advance characteristic of the higher class being appreciable only by a comparison of it with the skeleton of the most batrachoid of fishes—*e. g.* the protopterus (No. 380).

We then perceive that the bodies of the vertebræ, in the true batrachian, are distinctly ossified, though preserving, in the perennibranchiate species, a deep, conical, jelly-filled cavity both before and behind (Cut 11), C; they

Fig. 11.

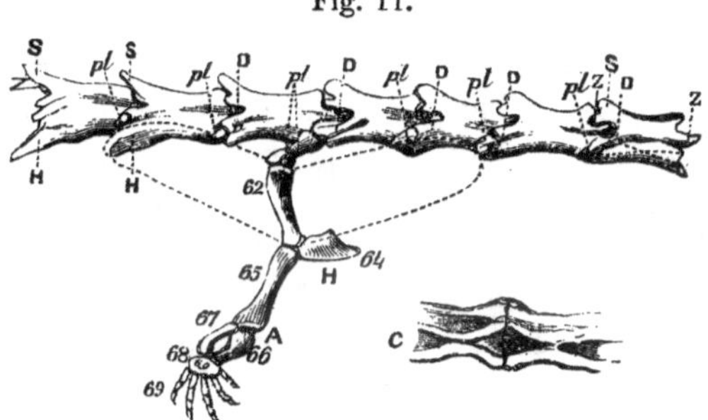

SACRAL VERTEBRA AND CONTIGUOUS VERTEBRÆ—MENOPOME.

have also coalesced with the neural arches, as these have with their spines, which are, however, scarcely prominent, except in the tail. The transverse processes are developed not only from the centrum but from the base of the neural arch, and are formed by both parapophyses and diapophyses; and they coexist with distinct hæmapophyses in the tail (*ib.*), H. With these, likewise, coexist cartilaginous pleurapophyses (*ib.*), *pl*, in the second, third, and fourth caudal vertebræ; short ossified pleurapophyses being developed from the ends of the diapophyses in the first caudal to the vertebra dentata inclusive.

By this instructive condition of the skeleton of the menopome, we perceive at once that the hæmapophyses (*ib.*), H, are neither transverse processes, nor ribs bent down or displaced, but are elements of vertebræ, as distinct as the neurapophyses above. The neural arches are now articulated together by well-developed zygapophyses with synovial articulations, which are absent in the protopterus, as in most fishes.

In the protopterus, as in the squatina and some other cartilaginous fishes, the neural arch of the atlas rests upon a backward production of the basioccipital; in the batrachians it is confluent with its own proper centrum, which developes two articular surfaces for the two occipital condyles. The hæmal arch of the occipital segment, which is attached to its proper vertebra in the protopterus (Fig. 32), A, 51, 52, as in osseous fishes, is detached and displaced backwards in the batrachians (Fig. 33), 51, 52. In the completion of the hæmal arch of the sacral vertebra in the menopome, by the enlargement of its transverse process (Fig. 11), D, and by its pleurapophysis (*ib.*), *pl*, extended to join a hæmapophysis (*ib.*), H, below, we have the key to the essential nature of the pelvis in all air-breathing animals. The progressive development of the appendages of the scapular and pelvic arches, which are to become the four limbs of air-breathing vertebrates, should be traced from their condition in the protopterus. Here (Fig. 32) they are reduced to a single ray, which is soft and many-jointed. In the *Amphiuma didactyla* (Fig. 33) the ray is ossified; its first joint (*ib.*), 53, is long, its second (*ib.*), 54, 55, is bifid, and a cartilage at the end of this supports two short terminal rays. This is the pattern of the subdivision of the appendage both of the scapular and pelvic arches, in all the higher vertebrates; hence, in consequence of the vast modifications of the several segments, the necessity for their special names. In the fore-limb the first segment (Fig. 33), 53, is the "arm," and its bone, the "humerus," No. 53; the second segment is the forearm—its two bones are the "radius," No. 55, and "ulna," No. 54; the third segment is the "hand"—its rays are the "fingers;" and its bones are subdivided into "carpals," No. 56, "metacarpals," and "phalanges," No. 57. In the hind-limb (Fig. 34), the *first* segment is the "thigh," and its bone, the "femur," No. 65; the second segment is the "leg," and its two bones are the "tibia," No. 66, and "fibula," No. 67; the third segment is the "foot"—its rays are the "toes;" its bones are subdivided into "tarsals," "metatarsals," and "phalanges."

In the siren, the pelvic arch and limbs are not developed; but they coexist with the scapular arch and limbs in all other batrachia. In the proteus, the last segment of the fore-limb divides into three rays, that of the hind-limb into two rays; in other words, it has three fingers and two toes. The menobranchus has four fingers and four toes. The axolotl has four fingers and five toes. The menopome has five fingers and five toes.

The ultimate subdivisions of the radiated or diverging appendages of the scapular and pelvic arches do not exceed five in any existing air-breathing animal, and their further complexity is due to the specialization of each digit, so as to combine in associated action, instead of their indefinite multiplication, which causes the seeming complexity of the same appendages in fishes.

In all the fish-like batrachia, called, from a retention of more or less of the branchial apparatus, "perennibranchia," the limbs are short, and the rays of the terminal segments of each limb are, more or less, united by a web; the body is long, and the tail long and compressed. But a great ascent in the scale of life is made in the batrachian order: all the species when hatched have the fish-like form, and gills for breathing water; most of them exist for some time, under this form, in water; and these undergo so strange a modification of form and structure before arriving at maturity, that it has been called a "metamorphosis." They change their aquatic for a terrestrial life; they breathe air instead of water; and from being omniverous become carnivorous. The tadpoles of our common toad and frog afford ready and abundant instances for tracing these stages. The following is an outline of the main phenomena of the change observable in regard to the osseous system:—

In the development of the skeleton of the common frog, a fibrous and cartilaginous framework is originally laid down conformably with the aquatic habits and life of the larva. A large cartilaginous cranium with four hæmal arches, and one of these supporting the framework of the branchial apparatus—a short series of fibro-cartilaginous vertebræ, minus the hæmal arches, in the trunk, and a series of fibrous septa diverging from the fibrous capsule of the notochord, and defining and giving attachment to the muscular segments along the tail—constitute the skeleton of the newly-hatched tadpole. As it grows, ossification begins; but only in those parts of the skeleton which are to be retained in the future frog. Thus, the centrums and neurapophyses of the head and trunk are ossified, but not those of the tail. In the trunk, ossification of the vertebral body proceeds centripetally by layers, successively diminishing in extent, and conical interspaces are left, consisting of the changed fibrous capsule of the notochord with the inclosed gelatinous cells, their liquefied contents forming the balls of fluid, between the biconcave vertebræ, as in fishes. But ossification proceeds to fill up the hinder cavity of the centrum, and to project into the front cavity of the succeeding vertebra, with which it is finally connected by a synovial ball-and-socket joint. Thus, the firmer intervertebral articulations are established, which adapt the vertebral column to the support of a body which is to be suspended upon limbs, and transported by them along the surface of the dry ground. Whilst this change is proceeding, the tail is undergoing rapid absorption, the retained fibro-cartilaginous condition of its vertebræ rendering them more ready for removal. In the last fused rudiments of the caudal vertebræ, ossification extends continuously, and the peculiar style (Fig. 12), *c*, at the end of the vertebral series in the frog and other tailless batrachians, is thus established.

In the conversion of the biconcave into cup-and-ball vertebræ in batrachian larvæ, ossification commonly, but not always, proceeds to obliterate the hinder cavity. In the land salamanders, however, it extends from the front cavity; so that in the adult vertebræ the ball is anterior, and the cup posterior, as in certain salamandroid fishes —*e. g.*, *lepidosteus*. In those batrachians that retain more or less of the branchial apparatus, with the outward form and natatory tail adapted to aquatic life, the vertebræ of the tail are ossified like those of the trunk, but the biconcave structure and intervening gelatinous joints are retained throughout life.

The chief changes which take place in the conversion of the cartilaginous skull of the larva to the ossified one

Fig. 12.

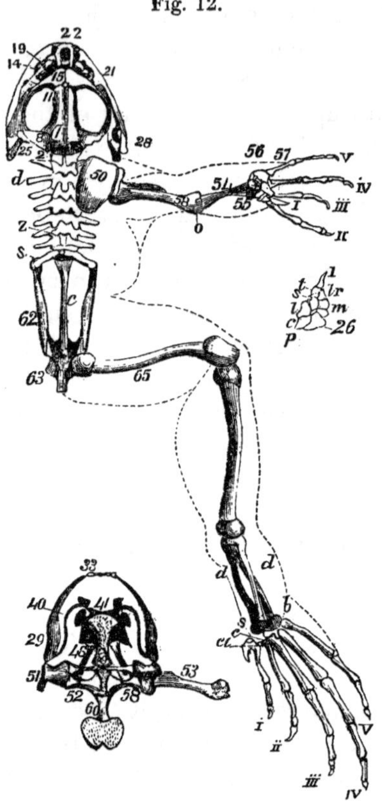

SKELETON OF THE FROG (*Rana esculenta*).

of the imago, or perfect frog, are seen in the shape and relative position of the hæmal arches and their appendages—*i. e.*, of the maxillary, mandibular, hyoid, and scapular arches. The maxillary arch expands in breadth, the mouth widens, and the horny mandibles are shed. As the mouth advances forwards, the tympanic pedicles are elongated, and are placed more obliquely; their proximal end retrograding from the post-frontal to the mastoid region of the skull, and their distal end inclining forwards with the attached lower jaw, Nos. 29, 33, on which the denticles now begin to be developed. For the still more extraordinary changes of the hyoid arch, No. 41, and its branchial appendages, No. 46, the student is referred to *Dugè's Recherches sur l'Ostéologie des Batraciens*, 4to., 1835; and to the writer's *Archetype of the Vertebrate Skeleton*, pp. 70, 71.

The scapular arch, which was close to the occiput, whilst protecting and supporting the branchial heart—its primary function—begins, as the rudiments of the fore-limbs bud out, to recede backwards, like the mandibular and branchial arches, but to a greater extent, the attachment to the occipital segment being wholly lóst. The scapular and coracoid portions of the arch become first ossified; the suprascapular plate remains long cartilaginous, and always partly so; the sternum is developed in proportion as the hyoid arch is reduced, and the branchial arches are removed; thus a strong fulcrum is completed for the articulation of the shoulder-joints. The pelvic arch had previously been completed, and the iliac bones and sides of the sacrum become coelongated: then the ilia continue to extend backwards as the tail is being absorbed, and the hind-limbs are lengthened out and finished.

Thus metamorphosed, the skeleton of the frog presents the following structure (Fig. 12): The number of vertebræ of the trunk, exclusive of the coxygeal style, *c*, is nine; the first, or atlas, has no diapophyses, but these are present and long on the rest, especially on the third, *d*, and ninth, *s*, vertebræ; in the latter they are thick, stand outwards, and support two other long, curved, rib-like bones, 62, which expand at their distal ends, and unite to two bony plates, 63, completing the hæmal arch of the ninth segment of the trunk. The bones of the hinder extremities are attached to the point of union of the above costal and hæmal pieces, one of which answers to the ilium, 62, and the other to the ischium, 63. The superior development of this arch relates to the great size and strength of the hinder extremities in the tailless tribe. The bodies of the vertebræ are articulated by ball-and-socket joints, the cup being anterior, the ball posterior, a modification which relates to the more terrestrial habits and locomotion of these higher-organized batrachia. The caudal vertebræ are represented by a single, elongated, cylindrical style, *c*, having an anchylosed neural canal. In the seven vertebræ, between the atlas and the sacrum, two zygapopophyses, looking upwards, two zygapophyses, *z*, looking downwards, and a short spine, are developed from each neural arch.

The suprascapula, 50, is very broad, and in great part ossified; the scapula, 51, divides at its humeral end into an acromial and coracoid process; the latter articulates with the true coracoid bone, 52, the acromion with the expanded extremity of the clavicle, 58: the glenoid cavity is formed by both the scapula and the coracoid. An episternal bone, 59, supporting a broad cartilage, is articulated to the mesial union of the clavicles, from which a bony bar is continued backwards between the expanded and partially conjoined ends of the coracoids. The sternum, 60, is articulated to the posterior part of the same extremities of the coracoids, and supports a broad "xiphoid" cartilage.

The proximal end of the humerus, 53, is an epiphysis; the distal end presents a hemispherical ball between a small external ridge, and a large internal condyloid process. The antibrachial bones have coalesced, but an anterior and posterior indentation at the distal half indicates the radius, 55, and ulna, 54; their distal articular extremities are represented by a single epiphysis. The ulnar portion of the bone develops a short and broad olecranon, *o*. The bones of the carpal series now receive definite names, and are as follows: (Fig. 12), *s*, scaphoid; *l*, lunare; *c* and *p*, cuneopisiforme; *t*, trapezium; *tr*, trapezoides; *m*, magnum; *u*, unciforme—here two distinct bones. The first digit, I, has one bone, a metacarpal; the second digit, II, has a metacarpal and two phalanges; the third, III, the same; the fourth, IV, has a metacarpal and three phalanges; and the fifth, V, the same.

Both the proximal and the distal extremities of the femur, 65, are in the condition of epiphyses. The tibia and fibula are connate, 66: a longitudinal impression on the front and back part of the expanded distal end indicates their division, but a single epiphysis, partially anchylosed, forms the proximal extremity, and a similar one the distal extremity, of the connate bones; they are perforated near their middle, from before backwards, by a vascular canal. The tarsal bones are now distinguished by names.

The astragalus, *a*, and calcaneum, *cl*, are much elongated; the former is slightly bent, the latter straight; they have coalesced at their proximal and also at their distal extremities with each other, and with the scaphoid, *s*, and

Fig. 13.

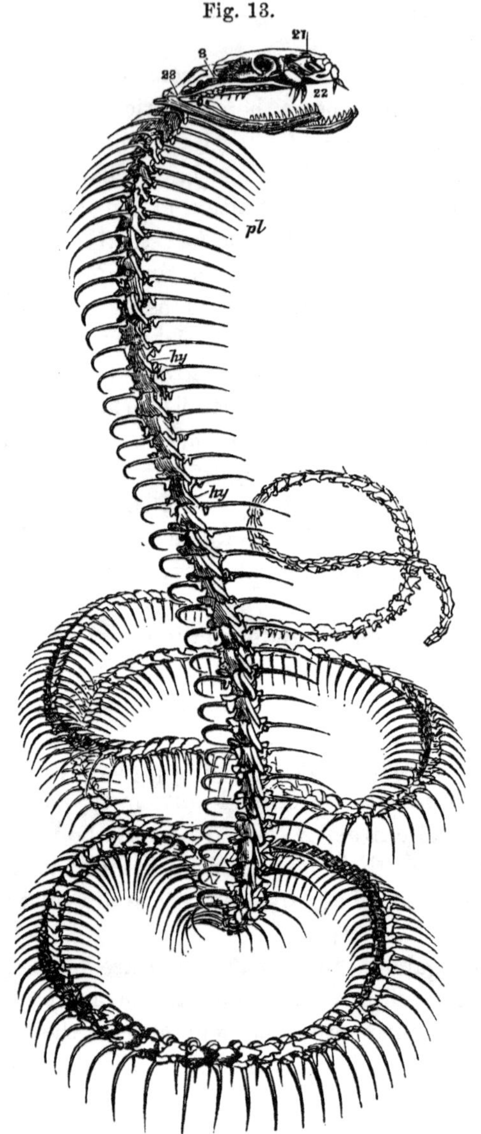

SKELETON OF THE COBRA (*Naija tripudians*).

cuboid, *b*, bones. Three cuneiform bones, *c*, *ci*, remain detached, and immediately support the three inner toes and a cartilaginous appendage. The first toe, *i*, and second toe, *ii*, have each a metatarsal and two phalanges; the third toe, *iii*, has a metatarsal and three phalanges; the fourth toe, *iv*, has a metatarsal and four phalanges; the fifth toe, *v*, a metatarsal and three phalanges. The great length and strengh of the pelvic arch, and its appendages, the hind-limbs, give the frog the power of executing the long leaps for which it is proverbial.

All the batrachia present this structure in common with fishes, viz., that the ribs of the trunk, when present, are free, consist only of "pleurapophyses," and do not encompass the thoracic-abdominal cavity. The absence of unyielding osseous girdles at this part seems to relate to a peculiarity of their generation, viz: the almost simultaneous ripening of the sperm-cells and ova, causing a great and sudden distension of the abdomen at the breeding period.

OSTEOLOGY OF THE OPHIDIA, OR SERPENT TRIBE.

There are certain tropical land batrachia—the Ceciliæ, *e. g.*—in which the body is as long and slender as in serpents, includes almost as numerous vertebræ, and is devoid of all trace of limbs. But the osteology of the typical Ophidian reptiles differs from that of the batrachians in the more elongated ribs; in the distinct basi- and superoccipitals; in the superoccipitals forming part of the ear-chamber; in the basioccipital combining with the exoccipital to form a single articular condyle for the atlas; in the ossification of the membranous space between the elongated parietals and the sphenoid; in the constant coalescence of the parietals with one another; in the constant confluence of the orbitosphenoids with the frontals, and in the meeting of the orbitosphenoids below the prosencephalon, upon the upper surface of the presphenoid; in the presence of distinct postfrontals, and the attachment thereto of the ectopterygoids, whereby they form an anterior point of suspension of the lower jaw, through the medium of the pterygoid and tympanic bones; lastly, in the connation of the prefrontals and lachrymals.

In studying the osteology of the head of the python, as the type of the Ophidian Order, by the aid of the following description, the student may compare the disarticulated skull, No. 628, with that of the large skeleton, No. 602, in the Museum, Royal College of Surgeons: the bones are numbered as here referred to.

The basioccipital, 1, is subdepressed, broadest anteriorly, subhexagonal; smooth and concave at the middle above, with a rough sutural tract on each side, and a hypapophysis below, produced into a recurved point. The hinder facet of the basioccipital is convex, forming the lower half of the occipital condyle, which is supported on a short peduncular prolongation. The basioccipital unites above and laterally with the exoccipitals and alisphenoids, and in front with the basisphenoid, upon which it rests obliquely, and it supports the medulla oblongata on its upper smooth surface.

The exoccipitals, 2, 2, are very irregular subtriangular bones; each is produced backwards into a peduncular process, supporting a moiety of the upper half of the occipital condyle. The outer and fore part of the exoccipital expands into the irregular base of the triangle:

it is perforated by a slit for the eighth pair of nerves; it articulates below with the basioccipital; it is excavated in front to lodge the petrosal cartilage, where it articulates with the alisphenoid; it unites above with the superoccipital. The superoccipital, 3, is of a subrhomboidal form, sends a spine from its upper and hinder surface, expands laterally into oblong processes, is notched anteriorly, and sends down two thin plates from its under surface, bounding on the mesial side the surface for the cerebellum, and by the outer side forming the inner and upper parts of the acoustic cavities. The superoccipital articulates below with the exoccipitals and alisphenoids, and in front with the parietal, by which it is overlapped in its whole extent. The occipital vertebra is as if it were sheathed in the expanded posterior outlet of the parietal one (Fig. 17), the centrum resting on the oblique surface of that in front, and the anterior base of the neural spine entering a cavity in and being overlapped by that of the preceding neural spine: the analogy of this kind of "emboitement" of the occipital in the parietal vertebra with the firm interlocking of the ordinary vertebræ of the trunk is very interesting: the end gained seems to be, chiefly, an extra protection of the epencephalon—the most important segment to life of all the primary divisions of the cerebrospinal axis. The thickness of its immediately protecting walls (formed by the basi, ex, and superoccipitals) is equal to that of the same vertebral elements in the human skull; but they are, moreover, composed of very firm and dense tissue throughout, having no diploë: the epencephalon also derives a further and equally thick bony covering from the basisphenoid and the parietals, the latter being overlapped by the mastoids, which form a third covering to the cerebellum.

The basisphenoid, 5, and presphenoid, 9, form a single bone, and the chief keel of the cranial superstructure. The posterior articular surface looks obliquely upwards and backwards, and supports that of the vertebral centrum behind, as the posterior ball of the ordinary vertebræ supports the oblique cup of the succeeding vertebræ; here, however, all motion is abrogated between the two vertebræ, and the coadapted surfaces are rough and sutural. The basisphenoid presents a smooth cerebral channel above for the mesencephalon, in front of which a deep depression (sella) sinks abruptly into the expanded part of the bone, and there bifurcates, each fork forming a short cul-de-sac in the substance of the bone.

The alisphenoids, 6, form the anterior half of the fenestra ovalis, which is completed by the exoccipitals; and in their two large perforations for the posterior divisions of the fifth pair of nerves, as well as in their relative size and position, the alisphenoids agree with those of the frog. Each alisphenoid is a thick suboval piece, with a tubercular process on its under and lateral part; it rests upon the basisphenoid and basioccipital, supports the posterior part of the parietal and a portion of the mastoid, 8, and unites anteriorly with the descending lateral plate of the parietal bone.

The parietal, 7, is a large and long, symmetrical, roof-shaped bone, with a median longitudinal crest along its upper surface, where the two originally distinct moieties have coalesced. It is narrowest posteriorly, where it overlaps the superoccipital, and is itself overlapped by the mastoid: it is convex at its middle part on each side of the sagittal spine, and is continued downwards and inwards, to rest immediately upon the basisphenoid. This part of the parietal seems to be formed by an extension of ossification along a membranous space, like that which permanently remains so in the frog, between the alisphenoid and orbitosphenoid: the mesencephalon and the chief part of the cerebral lobes are protected by this unusually developed spine of the mesencephalic vertebra. The optic foramina are conjugational ones, between the anterior border of the lateral plate of the parietal and the posterior border of the corresponding plate of the frontal.

The frontals, 11, rest by descending lateral plates, representing connate orbitosphenoids, 12, upon the attenuated, pointed prolongation of the basisphenoid: the upper surface of each frontal is flat, subquadrate, broader than long in the boa, and the reverse in the python, where the roof of the orbit is continued outwards by a detached superorbital bone: there is a distinct, oval, articular surface near the anterior median angle of each frontal to which the prefrontal is attached: the angle itself is slightly produced, to form the articular process for the nasal bones. The smooth orbitosphenoid plate of the frontal joins the outer margin of the upper surface of the frontal at an acute angle; the inner side of each frontal is deeply excavated for the prolongation of the cerebral lobes, and the cavity is converted into a canal by a median vertical plate of bone at the inner and anterior end of the frontal. The frontals join the parietals and postfrontals behind, and, by the anchylosed orbital plates, the presphenoid below, the prefrontals and nasals before, and the superorbitals at their lateral margins. The orbitosphenoids have their bases extended inwards, and meet below the prosencephalon and above the presphenoid, as the neurapophyses of the atlas meet each other above the centrum. The anterior third part of such inwardly-produced base is met by a downward production of the mesial margin of the frontal, forming a septum between the olfactory prolongations of the brain, but is not confluent with the frontal bone: the outer portion of the orbitosphenoids ascends obliquely outwards, and is confluent with the under part of the frontal; it is smooth externally, and deeply notched posteriorly for the optic foramen.

The post-frontal, 12, is a moderately long trihedral bone, articulated by its expanded cranial end to the frontal and parietal, and bent down to rest upon the outer and fore angle of the ectopterygoid. It does not reach that bone in the boa, nor in poisonous serpents. In both the boa and python, it receives the anterior sharp angle of the parietal in a notch.

The natural segment which terminates the cranium anteriorly, and is formed by the vomerine, prefrontal, and nasal bones, is very distinct in the ophidians.

The vomer, 13, is divided, as in salamandroid fishes and batrachians, but is edentulous: each half is a long, narrow plate, smooth and convex below, concave above, with the inner margin slightly raised; pointed anteriorly, and with two processes, and an intervening notch above the base of the pointed end. The prefrontals, 14, are connate with the lachrymals, 73. The two bones which intervene between the vomerine and nasal bones are the turbinals, 19; they are bent longitudinally outwards in the form of a semicylinder about the termination of the olfactory nerves.

The spine of the nasal vertebra is divided symmetrically, as in the frog, forming the nasal bones, 15; they are elongated, bent plates, with the shorter upper part arching outwards and downwards, completing the olfactory canal above, and with a longer median plate, forming a vertical wall, applied closely to its fellow, except in front, where the nasal process of the premaxillary is received in the interspace of the nasals.

The acoustic capsule remains in great part cartilaginous; there is no detached centre of ossification in it; to whatever extent this capsule is ossified, it is by a continuous extension from the alisphenoid. The sclerotic capsule of the eye is chiefly fibrous, with a thin inner layer of cartilage; the olfactory capsule is in a great measure ossified, as above described.

MAXILLARY ARCH.

The palatine, 20, or first piece of this arch, is a strong, oblong bone, having the inner side of its obtuse anterior end applied to the sides of the prefontals and turbinals, and, near its posterior end, sending a short, thick process upwards and inwards for ligamentous attachment to the lachrymal, and a second similar process outwards as the point of suspension of the maxillary bone. Between these processes the palatine is perforated, and behind them it terminates in a point. The chief part of the maxillary, 21 (Fig. 14), is continued forwards from its point of suspension, increasing in depth, and terminating obtusely; a shorter process is also, as usual, continued backwards, and terminates in a point. The point of suspension of the maxillary forms a short, narrow, palatine process. A space occupied by elastic ligament intervenes between the maxillary and the premaxillary, 22, which is single and symmetrical, and

Fig. 14.

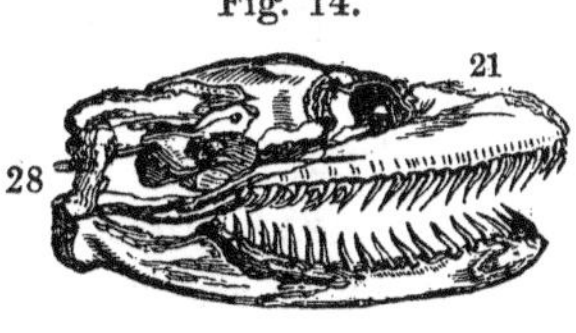

SKULL OF BOA CONSTRICTOR.

firmly wedged into the nasal interspace; the anterior expanded part of this small triangular bone supports two teeth. Thus, the bony maxillary arch is interrupted by two ligamentous intervals at the sides of the premaxillary key-bone, in functional relation to the peculiar independent movements of the maxillary and palatine bones required by serpents during the act of engulfing their usually large prey. Two bones extend backwards as appendages to the maxillary arch: one is the "pterygoid," 24, from the palatine; the other the ectopterygoid, 25, from the maxillary. The pterygoid is continued from the posterior extremity of the palatine to abut against the end of the tympanic pedicle; the under part of the anterior half of the pterygoid is beset with teeth. The ectopterygoid, 25, overlaps the posterior end of the maxillary, and is articulated by its posterior-obliquely cut end to the outer surface of the middle expanded part of the pterygoid.

MANDIBULAR ARCH.

The tympanic bone, 28 (Figs. 14 and 15), is a strong trihedral pedicle, articulated by an oblique upper surface to the end of the mastoid, and expanded transversely below to form the antero-posteriorly convex, transversely concave, condyle for the lower jaw. This consists chiefly of an articular and a dentary, with a small coronoid and splenial, piece. The articular piece ends obtusely, immediately behind the condyle; it is a little contracted in front of it, and gradually expands to its middle part, sends up two short processes, then suddenly contracts and terminates in a point wedged into the posterior and outer notch of the dentary piece. The articular is deeply grooved above, and produced into a ridge below. The coronoid is a short compressed plate; the splenial is a longer, slender plate, applied to the inner side of the articular and dentary, and closing the groove on the inner side of the latter. The outer side of the dentary offers a single perforation near its anterior end, which is united to that of the opposite ramus by elastic ligament.

By the above-described mode of union of the extremities of the maxillary and mandibular bones, those on the right side can be drawn apart from those on the left, and the mouth can be opened not only vertically, as in other vertebrate animals, but also transversely, as in insects. Viewing the bones of the mouth that support teeth in the great constricting serpents, they offer the appearance of six jaws—four above and two below; the inner pair of jaws above are formed by the palatine and pterygoid bones, the outer pair by the maxillaries, the under pair by the mandibles, or "rami," as they are termed, of the lower jaw.

Each of these six jaws, moreover, besides the movements vertically and laterally, can be protruded and retracted, independently of the other: by these movements the boa is enabled to retain and slowly engulf its prey, which may be much larger than its own body. At the first seizure, the head of the prey is held firmly by the long and sharp recurved teeth of all the jaws, whilst the body is crushed by the overlapping coils of the serpent; the death-struggles having ceased, the constrictor slowly uncoils, and the head of the prey is bedewed with an abundant slimy mucus: one jaw is then unfixed, and its teeth withdrawn by being pushed forward, when they are again infixed, further back upon the prey; the next jaw is then unfixed, protruded, and reattached; and so with the rest in succession—this movement of protraction being almost the only one of which they are susceptible whilst stretched apart to the utmost by the bulk of the animal encompassed by them; thus, by their successive movements, it is slowly and spirally introduced into the wide gullet.

Fig. 15.

SKULL OF A POISONOUS SNAKE.

The bones of the mouth, in the poisonous serpents, have characters distinct from those of the constricting serpents. These characters consist chiefly in the modification of form and attachments of the superior maxillary bone (Fig. 15), 21, which is movably articulated to the palatine, ectopterygoid, and lachrymal bones; but chiefly supported by the latter; which presents the form of a short, strong, three-sided pedicle, extending from the anterior external angle of the frontal to the anterior and upper part of the maxillary. The articular surface of the maxillary is slightly concave, of an oval shape; the surface articulating with the ectopterygoid on the posterior and upper part of the maxillary is smaller and convex. The maxillary bone is pushed forward and rotated upon the lachrymal joint by the advance of the ectopterygoids, which are associated with the movements of the tympanic pedicle of the lower jaw by means of the true pterygoid bones. The premaxillary bone (Fig. 13), 22, is edentulous. A single, long, perforated poison-fang is anchylosed to the right maxillary, and sometimes two similar fangs, as in the cobra figured in Cut 13. The palatine bones have four or five, and the pterygoids from eight to ten small, imperforate, pointed, and recurved teeth. The frontal bones are broader than they are long; there are no superorbitals. A strong ridge is developed from the under surface of the basisphenoid, and a long and strong recurved hypapophysis from that of the basioccipital; these give insertion to the powerful "longi-colli" muscles, by which the downward stroke of the head is performed in the infliction of the wound by the poison-fangs.

The characteristics of the trunk-vertebræ of the ophidian reptiles are as follows: The autogenous elements, except the pleurapophyses (Fig. 16), *pl*, coalesce with one another in the vertebræ of the trunk; and the pleurapophyses also become anchylosed to the diapophyses in those of the tail. There is no trace of suture between the neural arch (*ib.*), *n*, and centrum, *c*. The outer substance of the vertebra is compact, with a smooth or polished surface. The vertebræ are "procœlian;" that is, they are articulated together by ball-and-socket joints, the socket being on the fore part of the centrum, where it forms a deep cup with its rim sharply defined; the cavity looking not directly forwards, but a little downwards, from the greater prominence of the upper border; the well-turned prominent ball terminates the back part of the centrum rather more obliquely, its aspect being backwards and upwards. The hypapophysis, *hy*, is developed in different proportions from different vertebræ, but throughout the greater part of the trunk presents a considerable size in the cobra and crotalus (Figs. 13 and 16), *hy*; it is shorter in the python and boa. A vascular canal perforates the under surface of the centrum, and there are sometimes

Fig. 16.

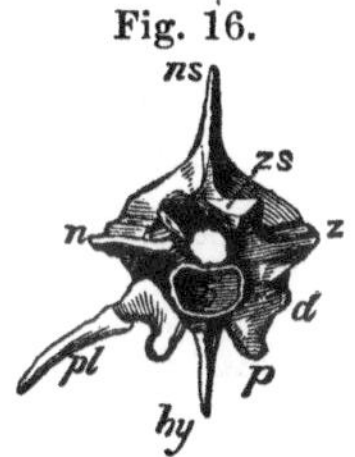

VERTEBRA OF THE RATTLESNAKE (*Crotalus*).

two or even three smaller foramina. In the python, a large, vertically oblong, but short diapophysis extends from the fore part of the side of the centrum obliquely backwards: it is covered by the articular surface for the rib, convex lengthwise, and convex vertically at its upper half, but slightly concave at its lower half. In the rattlesnake, the diapophysis develops a small, circumscribed, articular tubercle, *d*, for the free vertebral rib or pleurapophysis, *pl*; a parapophysis, *p*, extends downwards and forwards below the level of the centrum; the anterior zygapophysis, *z*, seems to be supported by a similar process from the upper end of the diapophysis. The base of the neural arch swells outward from its confluence with the centrum, and develops from each angle a transversely-elongated zygapophysis; that from the anterior angle looking upwards, that from the posterior angle downwards, both surfaces being flat, and almost horizontal, as in the batrachians. The neural canal is narrow; the neural spine, *ns*, is of moderate height, about equal to its antero-posterior extent; it is compressed and truncate. A wedge-shaped process (the "zygosphene"), *zs*, is developed from the fore part of the base of the spine; the lower apex of the wedge being, as it were, cut off, and its sloping sides presenting two smooth, flat, articular surfaces. This wedge is received into a cavity (the "zygantrum") excavated in the posterior expansion of the neural arch, and having two smooth articular surfaces to which the zygosphenal surfaces are adapted.

Thus the vertebræ of serpents articulate with each other by eight joints in addition to those of the cup and ball on the centrum; and interlock by parts reciprocally receiving and entering one another, like the joints called tenon-and-mortise in carpentry. In the caudal vertebræ, the hypapophysis is double, the transition being effected by its progressive bifurcation in the posterior abdominal vertebræ. The diapophyses become much longer in the caudal vertebræ, and support in the anterior ones short ribs which usually become anchylosed to their extremities.

The pleurapophyses or vertebral ribs in serpents have an oblong articular surface, concave above and almost flat below in the python, with a tubercle developed from the upper part, and a rough surface excavated on the fore part of the expanded head for the insertion of the precostal ligament. They have a large medullary cavity, with dense but thin walls, and a fine cancellous structure at their articular ends. Their lower end supports a short cartilaginous hæmapophysis, which is attached to the broad and stiff abdominal scute. These scutes, alternately raised and depressed by muscles attached to the ribs and integument, aid in the gliding movements of serpents; and the ribs, like the legs in the centipede, subserve locomotion; but they have also accessory functions in relation to breathing and constriction. The anterior ribs in the cobra (Fig. 13), *pl*, are unusually long, and are slightly bent; they can be folded back one upon another, and can be drawn forward, or erected, when they sustain a fold of integument, peculiarly colored in some species—*e. g.* the spectacled cobra—and which has the effect of making this venomous snake more conspicuous at the moment when it is about to inflict its deadly bite. The ribs commence in the cobra, as in other serpents, at the third vertebra from the head.

The centrum of the first vertebra coalesces with that of the second, and its place is taken by an autogenous hypapophysis: this, in the python, is articulated by suture to the neurapophyses; it also presents a concave articular surface anteriorly for the lower part of the basioccipital tubercle, and a similar surface behind for the detached central part of the body of the atlas, or "odontoid process of the axis." The base of each neurapophysis has an antero-internal articular surface for the exoccipital tubercle, the middle one for the hyapophysis, and a postero-internal surface for the upper and lateral parts of the odontoid; they thus rest on both the separated parts of their proper centrum. The neurapophyses expand and arch over the neural canal, but meet without coalescing. There is no neural spine. Each neurapophysis develops from its upper and hinder border a short zygapophysis, and from its side a still shorter diapophysis. In the second vertebra, the odontoid presents a convex tubercle anteriorly, which fills up the articular cavity in the atlas for the occipital tubercle; below this is the surface for the hypapophysial part of the atlas, and above and behind it are the two surfaces for the atlantal neurapophyses. The whole posterior surface of the odontoid is anchylosed to the proper centrum of the axis, and in part to its hypapophysis. The neural arch of the axis develops a short ribless diapophysis from each side of its base; a thick sub-bifid zygapophysis from each side of the posterior margin; and a moderately long bent-back spine from its upper part. The centrum terminates in a ball behind, and below this sends downwards and backwards a long hyapophysis.

At the opposite extreme of the elongated body, two or three much simplified vertebræ are usually found blended together. In true serpents there are no scapular arch and appendages, no sternum, no sacrum; but a pair of slender bones, often supporting a second bone, armed with a claw, are found suspended on the flesh near the vent. The exposed parts of these appendages are called "anal hooks;" the parts themselves, like the similarly suspended ventral fins of the pike, are rudiments of hind limbs.

Serpents have been regarded as animals degraded from a higher type; but their whole organization, and especially their bony structure, demonstrate that their parts are as exquisitely adjusted to the form of their whole, and to their habits and sphere of life, as is the organization of any animal which we call superior to them. It is true that the serpent has no limbs, yet it can outclimb the monkey, outswim the fish, outleap the jerboa, and, suddenly loosing the close coils of its crouching spiral, it can spring into the air and seize the bird upon the wing: all these creatures have been observed to fall its prey. The serpent has neither hands nor talons, yet it can outwrestle the athlete and crush the tiger in the embrace of its ponderous overlapping folds. Instead of licking up its food as it glides along, the serpent uplifts its crushed prey, and presents it, grasped in the death-coil as in a hand, to its slimy gaping mouth.

It is truly wonderful to see the work of hands, feet, and fins performed by a modification of the vertebral column—by a multiplication of its segments with mobility of its ribs. But the vertebræ are specially modified, as we have seen, to compensate, by the strength of their numerous articulations, for the weakness of their manifold repetition, and the consequent elongation of the slender column. As serpents move chiefly on the surface of the earth, their danger is greatest from pressure and blows from above; all the joints are fashioned accordingly to resist yielding, and sustain pressure in a vertical direction; there is no natural undulation of the body upwards and downwards—it is permitted only from side to side. So closely and compactly do the ten pairs of joints between each of the two hundred or three hundred vertebræ fit together, that even in the relaxed and dead state the body cannot be twisted except in a series of side coils. In the construction of the skull, which has merited a description in some detail, and well deserves a close study, the thickness and density of the cranial bones must strike the mind as a special provision against fracture and injury to the brain. When we contemplate the still more remarkable manner in which these bones are applied, one over another, the superoccipital (Fig. 17), 3, overlapping

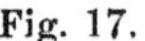

Fig. 17.

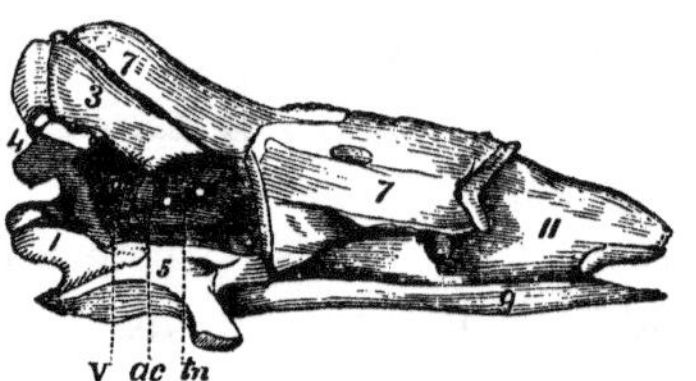

SECTION OF SKULL, BOA CONSTRICTOR.

the exoccipital, 4, and the parietal, 7, overlapping the superoccipital—the natural segments or vertebræ of the cranium being sheathed, one within the other, like the corresponding segments in the trunk—we cannot but discern a special adaptation in the structure of serpents to their commonly prone position, and a provision, exemplified in such structure, of the dangers to which they would be subject from falling bodies and the tread of heavy beasts. Many other equally beautiful instances of design might be cited from the organization of serpents, in relation to the necessities of their apodal vermiform character; just as the snake-like eel is compensated by analogous modifications amongst fishes, and the snake-like centipede among insects.

OSTEOLOGY OF LIZARDS.

The transition from the ophidian, or snake-like, to the lacertian, or lizard-like reptiles, is very gradual and easy, if we pass from the serpents with fixed jaws and a scapular arch—as, *e. g.* the slow-worms (*anguis*)—to the serpentiform lizards with mere rudiments of limbs—as, *e. g.* the pseudopus. The distinction is effected through the establishment of a costal arch in the trunk, completed by the addition of a hæmal spine (sternum) and hæmapophyses (sternal ribs) to the pleurapophyses or vertebral ribs, which are alone ossified in ophidia.

The vertebræ of the trunk have the same procœlian character, *i. e.* with the cup anterior and the ball behind; the latter being usually less prominent, more oblique, and more transversely oval than in serpents. The vertebræ also are commonly larger, and always fewer in number than in the typical ophidia. The ribs do not begin to be developed so near the head in lizards. Not only the atlas and dentata, but sometimes, as in the monitor (*varanus*), the four following vertebræ are devoid of pleurapophyses; and when these first appear they are short, and sometimes (as in *cyclodus*) expanded at their extremities. They rapidly elongate in succeeding vertebræ, and usually at the ninth from the head (*cyclodus*, *iguana*), or tenth (*varanus*), they are joined through the medium of ossified hæmopophyses to the sternum; two (*varanus*), three (*chameleo*, *iguana*), or four (*cyclodus*), following vertebræ are similarly completed; and then the hæmapophyses are either united below without intervening sternum (*chameleo*), or two or three of them are joined by a common cartilage to the cartilaginous end of the sternum. The hæmapophyses afterwards project freely, and are reduced to short appendages to the pleurapophyses. These also shorten, and sometimes suddenly, as, *e. g.* after the eighteenth vertebra in the monitors (*varanus*), in which they end at the twenty-eighth vertebra, as they began, viz: in the form of short straight appendages to the diapophyses.

The flying lizard (*Draco volans*), is so called on account of the wing-like expansions from the side of its body, supported, like the hood of the cobra, by slender elongated ribs. In this little lizard there are twenty vertebræ supporting movable ribs, which commence apparently at the fifth. Those of the eighth vertebra first join the sternum, as do those of the ninth and tenth; the pleurapophyses of the eleventh vertebra suddenly acquire extreme length; those of the five following vertebræ are also long and slender; they extend outwards and backwards and support the parachute formed by the broad lateral fold of the abdominal integuments. The pleurapophyses of the seventeenth vertebra become suddenly shorter, and these elements progressively diminish to the sacrum; this consists of two vertebræ, modified as in other lizards. There are about fifty caudal vertebræ.

The semi-ossified sternum in the iguana has a median groove and fissure, and readily separates into two lateral moieties. The long stem of the episternum covers the outer part of the groove, where it represents the *keel* of the sternum in birds.

In the skull of the lizard order we first meet with a second bony bar, diverging from the maxillary arch backwards, and abutting against the mastoid, and sometimes also against the tympanic and postfrontal. This bar is called the "zygomatic arch;" it usually consists of two bones—the one next the maxillary is the "malar," 26, the one next the mastoid is the "squamosal," 27; it assumes a form meriting that name in the tortoise, and first received it, as "pars squamosa," in man, where it is not only like a great scale, but becomes confluent with both the mastoid and tympanic. But, as has been before remarked, we must use the terms invented by anthropotomists as arbitrary signs of the corresponding bones in the lower creation.

The scapula in the monitor (*varanus*) is a triangular plate with a convex base, a concave hind border, and a nearly straight front border; the apex is thick and truncate, with an oval surface divided into two facets. The hind border forms a part of the glenoid cavity; the front one is a rough epiphysial surface, continuous with a similar but narrower tract, extending upon the anterior border, and by which the scapula articulates with the coracoid. In the iguanians and scincoids this synchondrosis is obliterated, and the two bones are confluent. The hind border of the scapula is nearly straight—the front one sends forwards a process dividing it into two deep marginations.

The coracoid in both the varanus and iguana is short and broad; its main body, which articulates with the sternum, is shaped like an axe-blade; and two strong, straight, compressed processes extend forwards from its neck, which is perforated between the origins of these processes and the part forming the glenoid articulations.

The clavicles are simple sigmoid styles in the varanus and iguana; are bent upon themselves, like the Australian boomerang, in the cyclodus; and have the median part of the bend expanded and perforated in lacerta and scincus. They are absent in the chameleon.

The sacral vertebræ retain, in some lacertians, the cup-and-ball joints; and in these, *e. g.* the scincoids—in which the centrums coalesce, the hind end of the second presents a ball to the first caudal—not a cup, as in the crocodile. In the cyclodus, the thick, short, straight pleurapophyses are distinct at their origins from the two coalesced centrums, but coalesce at their ends, that of the first sacral being the thickest. In varanus and iguana, the pleurapophyses as well as the centrums, retain their distinctness, but the hinder ribs incline forwards and touch the expanded ends of the fore pair. These ends are very thick, and are scooped out obliquely behind, so as to present a curved border to the ilium, which Cuvier compares to a horseshoe.

In the varanus and iguana, the pleurapophyses of the first caudal incline backwards as much as those of the second sacral do forwards. In the cyclodus they extend outwards, parallel with those of the sacral vertebræ, and are longitudinally grooved beneath. Hæmapophyses are wanting in the first caudal, are developed in the second, and are displaced to the interval between this and the third; they are confluent at their distal end, and produced into a long spine. At the twelfth tail-vertebra, the line is obvious that indicates the extent of the anterior detached piece, or epyphysis, of the centrum, immediately in front of the origin of the diapophyses; it continues marking off the anterior third of the centrum in all the other caudals. At this line the tail snaps off, when a lizard escapes by the common ruse of leaving the part of the tail by which it had been seized in the hands of the baffled pursuer. It is a very curious character, and quite peculiar to the lacertians—this ossification of the centrum from two points and their incomplete coalescence: it adds nothing to the power of bending, or to any other action of the tail, but indicates a prevision of the liability to their being caught by their long tail, and may be interpreted as a provision for their escape. The neural arch has coalesced with the centrum throughout the tail; the epiphysial line does not extend through that arch; but its thin and brittle walls soon break, when the two parts of the centrum are forcibly separated.

Lizards, as is well known, have the power of reproducing the tail, but the vertebral axis is never ossified in the new-formed part.

OSTEOLOGY OF CROCODILES.

The numerous and varied forms of fossil bones of extinct reptiles derive most elucidation from the skeleton of the higher organized sauria of Cuvier, which now are rightly held to constitute a distinct order, called *Loricata*, or *Crocodilia;* a more complete description, therefore, will be given of the skeleton of a member of this order, than was deemed needful in regard to the lacertian group of sauria.

Fig. 18.

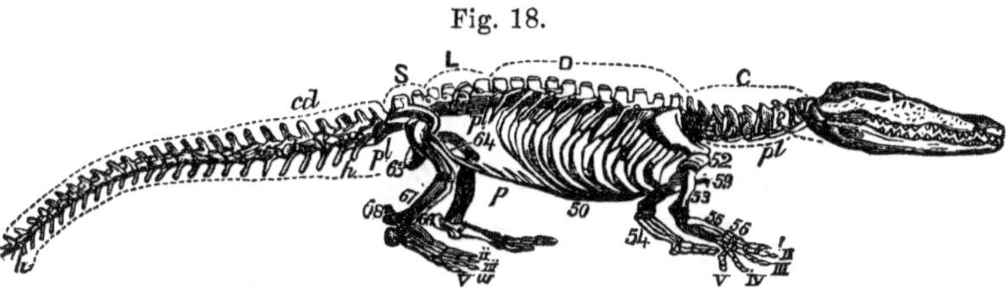

SKELETON OF THE CROCODILE (*Crocodilus niloticus*).

Commencing with the trunk, the first and second vertebræ of the neck are peculiarly modified in most air-breathing vertebrata, and have accordingly received the special names, the one of "atlas," the other of "axis." In comparative anatomy these become arbitrary terms, the properties being soon lost which suggested those names to the human anatomist; the "atlas," *e. g.*, has no power of rotation upon the "axis," in the crocodile, and it is only in the upright skeleton of man that the large globular head is sustained upon the shoulder-like processes of the "atlas." In the crocodile, these vertebræ are concealed by the peculiarly prolonged angle of the lower jaw in the side view of the skeleton (Fig. 18), and a figure of the two vertebræ is therefore subjoined (Fig. 19). The pleurapophyses, *pl*, are retained in both segments, as in all the other vertebræ of the trunk. That of the atlas, *pl, a*, is a simple slender style, articulated by the head only, to the "hypapophysis," *ahy*. The neurapophyses, *na*, of the atlas retain their primitive distinctness; each rests in part upon the proper body of the atlas, *ca*, in part upon the hypapophysis. The neural spine, *ns, a*, is also here an independent part, and rests upon the upper extremities of the neurapophyses. It is broad and flat, and prepares us for the further metamorphosis of the corresponding element in the cranial vertebræ.

Fig. 19.

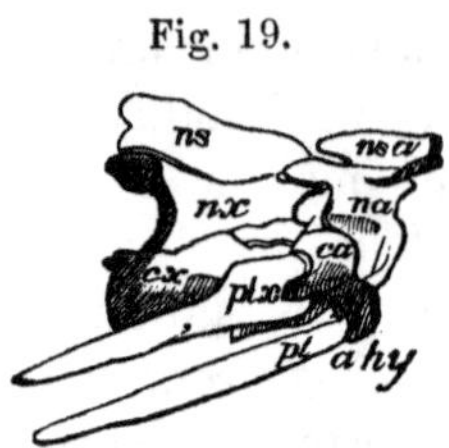

ATLAS AND AXIS VERTEBRÆ OF THE CROCODILE.

The centrum of the atlas, *ca*, called the "odontoid process of the axis" in human anatomy, here supports the abnormally-advanced rib of the axis vertebra, *pl, x*. The proper centrum of the axis vertebra, *cx*, is the only one in the cervical series which does not support a rib; it articulates by suture with its neurapophyses *nx*, and is characterized by having its anterior surface flat, and its posterior one convex.

With the exception of the two sacral vertebræ, the bodies of which have one articular surface flat and the other concave, and of the first caudal vertebra, the body of which has both articular surfaces convex, the bodies of all the vertebræ beyond the axis have the anterior articular surface concave, and the posterior one convex, and articulate with one another by ball-and-socket joints. This type of vertebra, which I have termed "procœlian" (προς, before, κοιλος, concave), characterizes all the existing genera and species of the order *Crocodilia* with all the extinct species of the tertiary periods, and also two extinct species of the greensand formation in New Jersey.[1] Here, so far as our present knowledge extends, the type was lost, and other dispositions of the articular surfaces of the centrum occur in the vertebræ of the crocodilia of the older secondary formations. The only known crocodilian genus of the periods antecedent to the chalk and greensand deposits with vertebræ articulated together by ball-and-socket joints, have the position of the cup and the ball the reverse of that in the modern crocodiles; and one genus, thus characterized by vertebræ of the "opisthocœlian" type (οπισθος, behind, κοιλος, concave), has accordingly been termed streptospondylus, signifying "vertebræ reversed." But the most prevalent type of vertebra amongst the crocodilia of the secondary periods was that in which both articular surfaces of the centrum were concave, but in a less degree than in the single concave sur-

[1] "Quarterly Journal of the Geological Society," November, 1849.

face of the vertebræ united by ball and socket. Vertebræ of this "amphicœlian" type (αμφι, both, κοιλος, concave) existed in the teleosaurus and steneosaurus. In the ichthyosaurus, the concave surfaces are usually deepened to the extent and in the form shown in those of the fish (Cut 8). Some of the most gigantic of the crocodilia of the secondary strata had one end of the vertebral centrum flattened, and the other (hinder) end concave; this "platycœlian" type (πλατυς, flat, κοιλος, concave) we find in the dorsal and caudal vertebræ of the gigantic cetiosaurus.

With a few exceptions, all the modern reptiles of the order lacertilia have the same procœlian type of vertebræ as the modern crocodilia, and the same structure prevailed as far back as the period of the mosasaurus, and in some smaller members of the lacertilian order in the cretaceous and wealden epochs.

Resuming the special description of the osteology of the modern crocodilia, we find the procœlian type of centrum established in the third cervical, which is shorter but broader than the second; a parapophysis is developed from the side of the centrum, and a diapophysis from the base of the neural arch; the pleurapophysis is shorter, its fixed extremity is bifid, articulating to the two abovenamed processes; its free extremity expands, and its anterior angle is directed forwards to abut against the inner surface of the extremity of the rib of both the axis and atlas, whilst its posterior prolongation overlaps the rib of the fourth vertebra. The same general characters and imbricated coadaptation of the ribs (Fig. 18), *pl*, characterize the succeeding cervical vertebræ to the seventh inclusive, the hypapophysis progressively though slightly increasing in size. In the eighth cervical the rib becomes elongated and slender; the anterior angle is almost or quite suppressed, and the posterior one more developed and produced more downwards, so as to form the body of the rib, which terminates, however, in a free point. In the ninth cervical, the rib is increased in length, but is still what would be termed a "false" or "floating rib" in anthropotomy.

In the succeeding vertebra the pleurapophysis articulates with a hæmapophysis, and the hæmal arch is completed by a hæmal spine; and by this completion of the typical segment we distinguish the commencement of the series of dorsal vertebræ (*ib.*), D. With regard to the so-called "perforation of the transverse process" this equally exists in the present vertebra, as in the cervicals; on the other hand, the cervical vertebræ equally show surfaces for the articulation of ribs. The typical characters of the segment, due to the completion of both neural and hæmal arches, are continued in some species of crocodilia to the sixteenth, in some (*crocodilus acutus*) to the eighteenth vertebra. In the *crocodilus acutus* and the *alligator lucius* the hæmapophysis of the eighth dorsal rib (seventeenth segment from the head) joins that of the antecedent vertebra. The pleurapophyses project freely outwards, and become "floating ribs" in the eighteenth, nineteenth, and twentieth vertebræ, in which they become rapidly shorter, and in the last appear as mere appendages to the end of the long and broad diapophyses: but the hæmapophyses by no means disappear after the solution of their union with their pleurapophyses; they are essentially independent elements of the segment, and they are continued, therefore, in pairs along the ventral surface of the abdomen of the crocodilia, as far as their modified homotypes the pubic bones. They are more or less ossified, and are generally divided into two or three pieces.

The lumbar vertebræ are those in which the diapophyses cease to support the movable pleurapophyses, although they are elongated by the coalesced rudiments of such which are distinct in the young crocodilia. The length and persistent individuality of more or fewer of these rudimental ribs determines the number of the dorsal and lumbar vertebræ respectively, and exemplifies the purely artificial character of the distinction. The number of vertebræ or segments between the skull and the sacrum, in all the crocodilia I have yet examined, is twenty-four. In the skeleton of a gavial, I have seen thirteen dorsal and two lumbar; in that of a crocodilus cataphractus, twelve dorsal and three lumbar; in those of a crocodilus acutus and alligator lucius, eleven dorsal and four lumbar, and this is the most common number; but in the skeleton of the crocodile, probably the species called croc. biporcatus, described by Cuvier, he gives five as the number of the lumbar vertebræ. But these varieties in the development or coalescence of the stunted pleurapophysis are of little essential moment; and only serve to show the artificial character of the "dorsal" and "lumbar" vertebræ. The coalescence of the rib with the diapophysis obliterates of course the character of the "costal articular surfaces," which we have seen to be common to both dorsal and cervical vertebræ. The lumbar zygapophyses have their articular surfaces almost horizontal, and the diapophyses, if not longer, have their antero-posterior extent somewhat increased; they are much depressed, or flattened horizontally.

The sacral vertebræ are very distinctly marked by the flatness of the coadapted ends of their centrums; there are never more than two such vertebræ in the crocodilia recent or extinct: in the first, the anterior surface of the centrum is concave; in the second, it is the posterior surface; the zygapophyses are not obliterated in either of these sacral vertebræ, so that the aspects of their articular surface—upwards in the anterior pair, downwards in the posterior pair—determines at once the corresponding extremity of a detached sacral vertebra. The thick and strong transverse processes form another characteristic of these vertebræ; for a long period the suture near their base remains to show how large a proportion is formed by the pleurapophysis. This element articulates more with the centrum than with the diapophysis developed from the neural arch; it terminates by a rough, truncate, expanded extremity, which almost or quite joins that of the similarly but more expanded rib of the other sacral vertebræ. Against these extremities is applied a supplementary costal piece, serially homologous with the appendage to the proper pleurapophysis in the dorsal vertebræ, but here interposing itself between the pleurapophyses and hæmapophyses of both sacral vertebræ, not of one only. This intermediate pleurapophysial appendage is called the "ilium;" it is short, thick, very broad, and subtriangular, the lower truncated apex forming with the connected extremities of the hæmapophysis an articular cavity for the diverging appendage, called the "hind leg." The hæmapophysis of the anterior sacral vertebra is called "pubis," 64; it is moderately long and slender, but expanded and flattened at its lower extremity, which is directed forwards towards that of its fellow, and joined to it through the intermedium of a broad, cartilaginous, hæmal spine, completing the hæmal canal. The posterior hæmopophysis, 63, is broader, subdepressed, and subtriangular, expanding as it approaches its fellow to complete the second hæmal arch; it is termed "ischium." The great development of all the elements of these hæmal arches, and the peculiar and distinctive forms of those that have thereby acquired, from the earliest dawn of anatomical science, special names, relates physiologically to the functions of the diverging appendage which is developed into a potent locomotive member. This limb appertains properly, as the proportion contributed by the ischium to the articular socket and the greater breadth of the pleurapophysis show, to the second sacral vertebra; to which the ilium chiefly belongs.

The first caudal vertebra, which presents a ball for articulating with a cup on the back part of the last sacral, retains, nevertheless, the typical position of the ball on the back part of the centrum; it is thus biconvex, and the only vertebra of the series which presents that structure.

The first caudal vertebra, moreover, is distinguished from the rest by having no articular surfaces for the hæmapophyses, which in the succeeding caudals form a hæmal arch, like the neurapophyses above, by articulating directly with the centrum. The arch so formed has its base not applied over the middle of a single centrum, but, like the neural arch in the back of the tortoise and sacrum of the bird, across the interspace between two centrums. The first hæmal arch of the tail belongs, however, to the second caudal vertebra, but it is displaced a little backwards from its typical position.

The caudal hæmapophyses, *h h*, coalesce at their lower or distal ends, from which a spinous process is prolonged downwards and backwards; this grows shorter towards the end of the tail, but is compressed and somewhat expanded antero-posteriorly. The hæmal arch so constituted has received the name of "chevron bone."

It is very true, as Cuvier said in the last lecture he delivered, "if we were agreed as to the crocodile's head, we should be so as to that of other animals; because the crocodile is intermediate between mammals, birds, and fishes." Accordingly, the following description of the crocodile's skull is coextensive with that of the fish; if the answerable bones are rightly determined between these, their correspondence with those of other vertebrates will be facilitated. The difficulties in comprehending the nature of some of the bones of the crocodile's head have arisen through passing to its comparison from that of the mammal's skull—by descending instead of ascending to it.

The segments composing the skull are more modified than those of the pelvis; but just as the vertebral pattern is best preserved in the neural arches of the pelvis, which are called collectively "sacrum," so, also, is it in the same arches of the skull, which are called collectively "cranium." The elements of which these cranial arches are composed, preserve, moreover, their primitive or normal individuality more completely than in any of the vertebræ of the trunk, except the atlas, and consequently the archetypal character can be more completely demonstrated.[1]

If, after separating the atlas from the occiput, we proceed to detach the occipital segment of the cranium from the next segment in advance, we find the detached segment presenting the form and structure of the neural arch. The "centrum" presents, like those of the trunk, a convexity or ball at its posterior articular surface, but its anterior one, like the hindmost centrum of the sacrum,

[1] The skull of the crocodile, partially disarticulated, and with the bones numbered as in the following description, may be had of Mr. Flower, No. 22 Lambeth Terrace, Lambeth Road.

unites with the next centrum in advance by a flat rough "sutural" surface. Like most of the centrums in the neck and beginning of the back, that of the occiput develops a hypapophysis, but this descending process is longer and larger, its base extending over the whole of the under surface of the centrum. It is a character whereby the occipital centrum of a crocodilian reptile may be distinguished from that of a lacertian one; for in the latter a pair of diverging hypapophyses project from the under surface, as is shown in most recent lizards and in the great extinct mosasaurus.

The upper and lateral parts of No. 1 present rough sutural surfaces, like those in the centrums of the trunk, for articulating with the "neurapophyses," Nos. 2, 2, which develop short, thick, obtuse, transverse processes, 4, 4. The modified or specialized character of the elements of the cranial vertebræ has gained for them special names. The centrum, 1, is called, as in fishes and all other vertebrates, the "basioccipital;" the neurapophyses, 2, 2, are the "exoccipitals;" the neural spine, 3, is the "superoccipital." The transverse processes, 4, 4, which may combine both diapophyses and paropophyses, are called the "paroccipitals;" they are never detached bones in the crocodilia, as they are in the chelonia and in most fishes. The exoccipitals perform the usual functions of neurapophyses, and, like those of the atlas, meet above the neural canal; they are perforated to give exit to the vagal and hypoglossal nerves, and protect the sides of the medulla oblongata and cerebellum—the two divisions of the epencephalon. The superoccipital, 3, is broad and flat, like the similarly detached neural spine of the atlas; it advances a little forwards, beyond its sustaining neurapophyses, to protect the upper surface of the cerebellum; it is traversed by tympanic air-cells, and assists with the exoccipitals, 2, 2, in the formation of the chamber for the internal ear.

The chief modification of the occipital segment of the skull, as compared with that of the osseous fish, or with the typical vertebra, is the absence of an attached hæmal arch. We shall afterwards see that this arch is present in the crocodile, although displaced backwards.

Proceeding with the neural arches of the crocodile's skull, if we dislocate the segment in advance of the occiput, we bring away, in connection with the long base-bone, 5, the bone, 9, which in the figure of the section of the serpent's skull (Cut 17) is shown similarly united to 5. In fact, the centrums of the vertebræ have here coalesced, as we find to happen in the neck of the siluroid fishes, and in the sacrum of birds and mammals. The two connate cranial centrums must be artificially divided, in order to obtain the segments distinct to which they belong. The hinder portion, 5, of the great base-bone, which is the centrum of the parietal vertebra, is called "basisphenoid." It supports that part of the "mesencephalon," which is formed by the lobe of the third ventricle, and its upper surface is excavated for the pituitary prolongation of that cavity. The basisphenoid develops from its under surface a "hypapophysis," which is suturally united with the fore part of that of the basioccipital, but extends further down, and is similarly united in front to the "pterygoids," 24. These rough sutural surfaces of the long descending process of the basisphenoid are very characteristic of that centrum, when detached, in a fossil state. The neurapophyses of the parietal vertebra, 6, 6, or the "alisphenoids," protect the sides of the mesencephalon, and are notched at their anterior margin, for a conjugational foramen transmitting the trigeminal nerve. As accessory functions they contribute, like the corresponding bones in fishes, to the formation of the ear-chamber. They have, however, a little retrograded in position, resting below in part upon the occipital centrum, and supporting more of the spine of that segment, 3, than of their own, 7. The spine of the parietal vertebra is a permanently distinct, single, depressed bone, like that of the occipital vertebra; it is called the "parietal," and completes the neural arch, as its crown or key-bone; it is partially excavated by the tympanic air-cells, and overlaps the superoccipital. The bones, 8, 8, wedged between 6 and 7, manifest more of their diapophysial character than their homotypes, 4, 4, do in the occipital segment, since they support modified ribs, are developed from independent centres, and preserve their individuality. They form no part of the inner walls of the cranium, but send outwards and backwards a strong transverse process for muscular attachment. They afford a ligamentous attachment to the hæmal arch of their own segment, and articulate largely with the pleurapophyses, 28, of the antecedent hæmal arch, whose more backward displacement, in comparison with its position in the fish's skull, is well illustrated in the metamorphosis of the toad and frog.

On removing the neural arch of the parietal vertebra, after the section of its confluent centrum, the elements of the corresponding arch of the frontal vertebra present the same arrangement. The compressed produced centrum has its form modified like that of the vertebral centrums at the opposite extreme of the body in many birds; it is called the "presphenoid." The neurapophyses, 10, 10, articulate with the upper part of 9; they are expanded, and smoothly excavated on their inner surface to support the sides of the large prosencephalon; they dismiss the great optic nerves by a notch. They show the same tendency to a retrograde change of position as the neighboring neurapophyses, 6; for though they support a greater proportion of their proper spine, 11, they also support part of the parietal spine, 7, and rest, in part, below upon the parietal centrum, 5: the neurapophyses, 10, 10, are called "orbitosphenoids." The neural spine, 11, of the frontal vertebra retains its normal character as a single symmetrical bone, like the parietal spine which it partly overlaps; it also completes the neural arch of its own segment, but is remarkably extended longitudinally forwards, where it is much thickened, and assists in forming the cavities for the eyeballs; it is called the "frontal" bone.

In contemplating in the skull itself, or such side view as is given in Fig. 9, p. 22, of my work on the *Archetype Skeleton*, the relative position of the frontal, 11, to the parietal, 7, and of this to the superoccipital, 3, which is overlapped by the parietal, just as itself overlaps the flattened spine of the atlas, we gain a conviction which cannot be shaken by any difference in their mode of ossification, by their median bipartition, or by their extreme expansion in other animals, that the above-named single, median, imbricated bones, each completing its neural arch, and permanently distinct from the piers of such arch, must repeat the same element in those successive arches—in other words, must be "homotypes," or serially homologous. In like manner the serial homology of those piers, called "neurapophyses," viz: the laminæ of the atlas, the exoccipitals, the alisphenoids, and the orbitosphenoids, is equally unmistakable. Nor can we shut out of view the same serial relationship of the paroccipitals, as coalesced diapophyses of the occipital vertebra, with the mastoids 8, and the postfrontals, 12, as permanently detached diapophyses of their respective vertebræ. All stand out from the sides of the cranium, as transverse processes for muscular attachment; all are alike autogenous in the turtles; and all of them, in fishes, offer articular surfaces for the ribs or hæmal arches of their respective vertebræ; and these characters are retained in the postfrontals as well as in the mastoids of the crocodiles.

The frontal diapophysis, 12, is wedged between the back part of the spine, 11, and the neurapophysis, 10; its outwardly projecting process extends also backwards, and joins that of the succeeding diapophysis, 8; but, notwithstanding the retrogradation of the inferior arch, it still articulates with part of its own pleurapophysial element, 28, which forms the proximal element of that arch.

There finally remain in the cranium of the crocodile, after the successive detachment of the foregoing arches, the bones terminating the forepart of the skull; but, notwithstanding the extreme degree of modification to which their extreme position subjects them, we can still trace in their arrangement a correspondence with the vertebrate type.

A long and slender symmetrical grooved bone, 13, between 24 and 24, like the ossified inferior half of the capsule of the notochord, is continued forwards from the inferior part of the centrum, 9, of the frontal vertebra, and stands in the relation of a centrum to the vertical plates of bone, 14, which expand as they rise into a broad, thick, triangular plate, with an exposed horizontal superior surface. These bones, which are called "prefontals," stand in the relation of "neurapophyses" to the rhinencephalic prolongations of the brain commonly but erroneously called "olfactory nerves;" and they form the piers or haunches of a neural arch, which is completed above by a pair of symmetrical bones, 15, called "nasals," which I regard as a divided or bifid neural spine.

The centrum of this arch is established by ossification in the expanded anterior prolongation of the fibrous capsule of the notochord, beyond the termination of its gelatinous axis. The median portion above specified retains most of the formal characters of the centrum; but there is a pair of long, slender, symmetrical ossicles, which, from the seat of their original development, and their relative position to the neural arch, must be regarded as also parts of its centrum. And this ossification of the element in question from different centres will be no new or strange character to those who recollect that the vertebral body in man and mammalia is developed from three centres. The term "vomer" is applied to the pair of bones, 13, because their special homology with the single median bone, so called in fishes and mammals, is indisputable; but a portion of the same element of the skull retains its single symmetrical character in the crocodile, and is connate with the enormous pterygoids, 24, between which it is wedged. In some alligators (*all. niger*) the divided anterior vomer extends far forwards, expands anteriorly, and appears upon the bony palate.

Almost all the other bones of the head of the crocodile are adjusted so as to constitute four inverted arches. These are the hæmal arches of the four segments or vertebræ, of which the neural arches have been just described. But they have been the seat of much greater modifications, by which they are made subservient to a variety of functions unknown in the hæmal arches of the rest of the body. Thus, the two anterior hæmal arches of the head perform the office of seizing and bruising the food; are armed for that purpose with teeth: and, whilst one arch is firmly fixed, the other works upon it like the hammer upon the anvil. The elements of the fixed arch, called "maxillary arch," have accordingly undergone the greatest amount of morphological change, in order to adapt that arch to its share in mastication, as well as for forming part of the passage for the respiratory medium, which is perpetually traversing this hæmal canal in its way to purify the blood. Almost the whole of the upper surface of the maxillary arch is firmly united to contiguous parts of the skull by rough or sutural surfaces, and its strength is increased by bony appendages, which diverge from it to abut against other parts of the skull. Comparative anatomy teaches that, of the numerous places of attachment, the one which connects the maxillary arch by its element, 20, with the centrum, 13, and the descending plates of the neurapophyses, 14, of the nasal segment, is the normal or the most constant point of its suspension, the bone, 20, being the pleurapophysial element of the maxillary arch: it is called the "palatine," because the under surface forms a portion of the bony roof of the mouth, called the "palate." It is articulated at its fore part with the bone, 21, in the same plates, which bone is the hæmapophysial element of the maxillary arch: it is called the "maxillary," and is greatly developed both in length and breadth; it is connected not only with 20 behind, and 22 in front, which are parts of the same arch, and with the diverging appendages of the arch, viz., 26, the malar bone, and 24, the pterygoid, but also with the nasals, 15, and the lachrymal, 16, as well as with its fellow of the opposite side of the arch. The smooth, expanded horizontal plate, which effects the latter junction, is called the palatal plate of the maxillary; the thickened external border, where this plate meets the external rough surface of the bone, and which is perforated for the lodgment of the teeth, is the "alveolar border" or "process" of the maxillary. The hæmal spine or key-bone of the arch, 22, is bifid, and the arch is completed by the symphysial junction of the two symmetrical halves; these halves are called "premaxillary bones:" these bones, like the maxillaries, have a rough facial plate, and a smooth palatal plate, with the connecting alveolar border. The median symphysis is perforated vertically through both plates; the outer or upper hole being the external nostril, the under or palatal one being the prepalatal or naso-palatal aperture.

Both the palatine and the maxillary bones send outwards and backwards parts or processes which diverge from the line of the hæmal arch, of which they are the chief elements; and these parts give attachment to distinct bones which form the "diverging appendages" of the arch, and serve to attach it, as do the diverging appendages of the thoracic hæmal arches in the bird, to the succeeding arch.

The appendage 24, called "pterygoid," effects a more extensive attachment, and is peculiarly developed in the crocodilia. As it extends backwards it expands, unites with its fellow below the nasal canal, and encompassing that canal, coalesces above it with the vomer, and is firmly attached by suture to the presphenoid and basisphenoid: it surrounds the hinder or palatal nostril, and, extending outwards, it gives attachment to a second bone, 25, called "ectopterygoid," which is firmly connected with the maxillary, 25, the malar, 26, and the postfrontal, 12. The second diverging ray is of great strength; it extends from the maxillary, 21 ("hæmapophysis" of the maxillary arch), to the tympanic, 28 ("pleurapophyses" of the mandibular arch), and is divided into two pieces, the malar, 26, and the squamosal, 27. Such are the chief crocodilian modifications of the hæmal arch, and appendages of the anterior or nasal vertebra of the skull.

The hæmal arch of the frontal vertebra is somewhat less metamorphosed, and has no diverging appendage. It is slightly displaced backwards, and is articulated by only a small proportion of its pleurapophysis, 28, to the parapophysis, 12, of its own segment; the major part of that short and strong rib articulating with the parapophysis, 8, of the succeeding segment. The bone, 28, called "tympanic," because it serves to support the "drum of the ear" in air-breathing vertebrates, is short, strong, and immovably wedged, in the crocodilia, between the paroccipital, 4, mastoid, 8, postfrontal, 12, and squamosal, 27; and the conditions of this fixation of the pleurapophysis are exemplified in the great development of the hæmapophysis (mandible), which is here unusually long, supports numerous teeth, and requires, therefore, a firm point of suspension, in the violent actions to which the jaws are put in retaining and overcoming the struggles of a powerful living prey. The movable articulation between the pleurapophysis, 28, and the rest of the hæmal arch is analogous to that which we find between the thoracic pleurapophysis and hæmapophysis in the ostrich and many other birds. But the hæmopophysis of the mandibular arch in the crocodiles is subdivided into several pieces, in order to combine the greatest elasticity and strength with a not excessive weight of bone. The different pieces of this purposely subdivided element have received definite names. That numbered 29, which offers the articular concavity to the convex condyle of the tympanic, 28, is called the "articular" piece; that beneath it, 30, which develops the angle of the jaw, when this projects, is the "angular" piece; the piece above, 29′, is the "surangular;" the thin, broad, flat piece, 31, applied, like a splint, to the inner side of the other parts of the mandible, is the "splenial;" the small accessory ossicle, 31′, is the "coronoid," because it develops the process, so called, in lizards; the anterior piece, 32, which supports the teeth, is called the "dentary." This latter is the homotype of the premaxillary, or it represents that bone in the mandibular arch, of which it may be regarded as the hæmal spine; the other pieces are subdivisions of the hæmapophysial element. The purport of this subdivision of the lower jaw-bone has been well explained by Conybeare[1] and Buckland,[2] by the analogy of its structure to that adopted in binding together several parallel plates of elastic wood or steel to make a crossbow, and also in setting together thin plates of steel in the springs of carriages. Dr. Buckland adds: "Those who have witnessed the shock given to the head of a crocodile by the act of snapping together its thin long jaws, must have seen how liable to fracture the lower jaw would be, were it composed of one bone only on each side." The same reasoning applies to the composite structure of the long tympanic pedicle in fishes. In each case the splicing and bracing together of thin flat bones of unequal length and of varying thickness, affords compensation for the weakness and risk of fracture that

[1] "Geol. Trans.," 1821, p. 565.

[2] "Bridgewater Treatise," 1836, vol. i. p. 176.

would otherwise have attended the elongation of the parts. In the abdomen of the crocodile, the analogous subdivision of the hæmapophyses, there called abdominal ribs, allows of a slight change of their length, in the expansion and contraction of the walls of that cavity; and since amphibious reptiles, when on land, rest the whole weight of the abdomen directly upon the ground, the necessity of the modification for diminished liability to fracture further appears. These analogies are important, as demonstrating that the general homology of the elements of a natural segment of the skeleton is not affected or obscured by their subdivision for a special end. Now this purposive modification of the hæmapophyses of the frontal vertebra, is but a repetition of that which affects the same elements in the abdominal vertebræ.

Passing next to the hæmal arch of the parietal vertebra, we are first struck by its small relative size. Its restricted functions have not required it to grow in proportion with the other arches, and it consequently retains much of its embryonic dimensions. It consists of a ligamentous "stylohyal," its pleurapophysis retaining the same primitive histological condition which obstructs the ordinary recognition of the same elements of the lumbar hæmal arches. A cartilaginous "epihyal," 39, intervenes between this and the ossified "hæmapophysis," 40, which bears the special name of ceratohyal. The hæmal spine, 41, retains its cartilaginous state, like its homotypes, in the abdomen; there they get the special name of "abdominal sternum," here of "basihyal." The basihyal has, however, coalesced with the thyrohyals to form a broad cartilaginous plate, the anterior border rising like a valve to close the fauces, and the posterior angles extending beyond and sustaining the thyroid and other parts of the larynx. The long, bony "ceratohyal," and the commonly cartilaginous "epihyal," are suspended by the ligamentous "stylohyal" to the paroccipital process; the whole arch having, like the mandibular one, retrograded from the connection it presents in fishes.

This retrogradation is still more considerable in the succeeding hæmal arch. In comparing the occipital segment of the crocodile's skeleton with that of the fish, the chief modification that distinguishes that segment in the crocodile is the apparent absence of its hæmal arch. We recognize, however, the special homologues of the constituents of that arch of the fish's skeleton in the bones 51 and 52 of the crocodile's skeleton (Fig. 18); but the upper or suprascapular piece, 50, retains, in connection with the loss of its proximal or cranial articulations, its cartilaginous state; the scapula, 51, is ossified, as is likewise the coracoid, 52, the lower end of which is separated from its fellow by the interposition of a median, symmetrical, partially-ossified piece called "episternum." The power of recognizing the special homologies of 50, 51, and 52 in the crocodile, with the similarly-numbered constituents of the same arch in fishes—though masked, not only by modifications of form and proportion, but even of very substance, as in the case of 50—depends upon the circumstance of these bones constituting the same essential element of the archetypal skeleton, viz., the fourth hæmal arch, numbered *pl*, 52, in Fig. 7: for although in the present instance there is superadded, to the adaptive modifications above cited, the rarer one of altered connections, Cuvier does not hesitate to give the same names, "suprascapulaire" to 50, and "scapulaire" to 51, in both fish and crocodile; but he did not perceive or admit that the narrower relations of special homology were a result of, and necessarily included in, the wider law of general homology. According to the latter law, we discern in 50 and 51 a compound "pleurapophysis," in 52 a "hæmapophysis," and in *hs*, the "hæmal spine," completing the hæmal arch.

The scapulo-coracoid arch, both elements, 51, 52, of which retain the form of strong and thick vertebral and sternal ribs in the crocodile, is applied in the skeleton of that animal over the anterior thoracic hæmal arches. Viewed as a more robust hæmal arch, it is obviously out of place in reference to the rest of its vertebral segment. If we seek to determine that segment by the mode in which we restore to their centrums the less displaced neural arches of the antecedent vertebræ of the cranium or in the sacrum of the bird,[1] we proceed to examine the vertebræ before and behind the displaced arch, with the view to discover the one which needs it, in order to be made typically complete. Finding no centrum and neural arch without its pleurapophyses from the scapula to the pelvis, we give up our search in that direction; and in the opposite direction we find no vertebra without its ribs, until we reach the occiput; there we have centrum and neural arch, with coalesced parapophyses, but without the hæmal arch, which arch can only be supplied by a restoration of the bones 50–52 to the place which they naturally occupy in the skeleton of the fish. And since anatomists are generally agreed to regard the bones 50–52 in the crocodile (Fig. 18) as specially homologous with those so numbered in the fish (Fig. 9), we must conclude that they are likewise homologous in a higher

[1] See "On the Archetype and Homologies of the Vertebrate Skeleton," pp. 117 and 159.

sense; that in the fish the scapula-coracoid arch is in its natural or typical position, whereas in the crocodile it has been displaced for a special purpose. Thus, agreeably with a general principle, we perceive that, as the lower vertebrate animal illustrates the closer adhesion to the archetype by the natural articulation of the scapulo-coracoid arch to the occiput, so the higher vertebrate manifests the superior influence of the antagonizing power of adaptive modification by the removal of that arch from its proper segment.

The anthropotomist, by this mode of counting and defining the dorsal vertebræ and ribs, admits, unconsciously perhaps, the important principle in general homology which is here exemplified; and which, pursued to its legitimate consequences, and further applied, demonstrates that the scapula is the modified rib of that centrum and neural arch, which he calls the "occipital bone;" and that the change of place which chiefly masks that relation (for a very elementary acquaintance with comparative anatomy shows how little mere form and proportion affect the homological characters of bones), differs only in extent, and not in kind, from the modification which makes a minor amount of comparative observation requisite, in order to determine the relation of the shifted dorsal rib to its proper centrum in the human skeleton.

With reference, therefore, to the occipital vertebra of the crocodile, if the comparatively well-developed and permanently-distinct ribs of all the cervical vertebræ prove the scapular arch to belong to none of those segments, and if that hæmal arch be required to complete the occipital segment, which it actually does complete in fishes, then the same conclusion must apply to the same arch in other animals, up to man himself.

The anterior locomotive extremity is the diverging appendage of the arch, under one of its numerous modes and grades of development. The proximal element of this appendage, or that nearest the arch, is called the "humerus," 53 (Fig. 18). The second segment of the limb consists of two bones; the larger one, 54, is called the "ulna:" it articulates with the outer condyle of the humerus by an oval facet, the thick convex border of which swells a little out behind, and forms a kind of rudimental "olecranon;" the distal end is much less than the proximal one, and is most produced at the radial side.

The radius, 55, has an oval head; its shaft is cylindrical; its distal end oblong and subcompressed.

The small bones, 56, which intervene between these and the row of five longer bones, are called "carpals;" they are four in number in the crocodilia. One seems to be a continuation of the radius, another of the ulna; these two are the principal carpals; they are compressed in the middle, and expanded at their two extremities: that on the radial side of the wrist is the largest. A third small ossicle projects slightly backwards from the proximal end of the ulnar metacarpal; it answers to the bone "pisiforme" in the human wrist. The fourth ossicle is interposed between the ulnar carpal and the metacarpals of the three ulnar digits.

These five terminal-jointed rays of the appendage are counted from the radial to the ulnar side, and have received special names; the first is called "pollex," the second "index," the third "medius," the fourth "annularis," and the fifth "minimus." The first joint of each digit is called "metacarpal;" the others are termed "phalanx." In the crocodilia, the pollex has two phalanges, the index three, the medius four, the annularis four, and the minimus three. The terminal phalanges, which are modified to support claws, are called "ungual" phalanges.

As the above-described bones of the scapular extremity are developments of the appendage of the scapular arch, which is the hæmal arch of the occipital vertebra, it follows, that, like the branchiostegal rays and opercular bones in fishes, they are essentially bones of the head. But the enumeration of the bones of the crocodile's skull is not completed by these; there is a bone anterior to the orbit, which is perforated at its orbital border by the duct of the lachrymal gland, whence it is termed the "lachrymal bone," and its facial part extends forwards between the bones marked 14, 15, 21, and 26. In many crocodilia there is a bone at the upper border of the orbit, which extends into the substance of the upper eyelid; it is called "superorbital." In the crocodilus palpebrosus there are two of these ossicles.

Both the lachrymal and superorbital bones answer to a series of bones found commonly in fishes, and called "suborbitals" and "superorbitals." The lachrymal is the most anterior of the suborbital series, and is the largest in fishes; it is also the most constant in the vertebrate series, and is grooved or perforated by a mucous duct. These ossicles appertain to the dermal or muco-dermal system or "exoskeleton," not to the vertebral system or "endoskeleton."

There remains, to complete this sketch of the osteology of the crocodile, a brief notice of the bones composing the diverging appendage of the pelvic arch: these being a repetition of the same element as the appendage of the scapular arch, modified and developed for a similar office, manifest a very close resemblance to it. The first bone, called the "femur," is longer than the humerus, and, like it, presents an enlargement of both extremities, with a double curvature of the intervening shaft, but the directions are the reverse of those of the humerus, as may be seen in Fig. 18, where the upper or proximal half of the femur is concave, and the distal half convex, anteriorly. The head of the femur is compressed from side to side, not from before backwards, as in the humerus; a pyramidal protuberance from the inner surface of its upper fourth represents a "trochanter;" the distal end is expanded transversely, and divided at its back part into two condyles. The next segment of the hind-limb or "leg," includes, like the corresponding segment of the fore-limb called "fore-arm," two bones. The largest of these is the "tibia," 66, and answers to the radius. It presents a large, triangular head to the femur; it terminates below by an oblique crescent with a convex surface. The "fibula" is much compressed above; its shaft is slender and cylindrical, its lower end is enlarged and triangular. The group of small bones which succeed those of the leg are the tarsals; they are four in number, and have each a special name. The "astragalus" articulates with the tibia, and supports the first and part of the second toe. The calcaneum intervenes between the fibula and the ossicle supporting the two outer toes; it has a short but strong posterior tuberosity. The ossicle referred to represents the bone called "cuboid" in the human tarsus. A smaller ossicle, wedged between the astragalus and the metatarsals of the second and third toes is the "ectocuneiform."

Four toes only are normally developed in the hind-foot of the crocodilia; the fifth is represented by a stunted rudiment of its metatarsal, which is articulated to the cuboid and to the base of the fourth metatarsal. The four normal metatarsals are much longer than the corresponding metacarpals. That of the first or innermost toe is the shortest and strongest; it supports two phalanges. The other three metatarsals are of nearly equal length, but progressively diminish in thickness from the second to the fourth. The second metatarsal supports three phalanges; the third four; and the fourth also has four phalanges, but does not support a claw. The fifth digit is represented by a rudiment of its metatarsal in the form of a flattened triangular plate of bone, attached to the outer side of the cuboid, and slightly curved at its pointed and prominent end.

The forms and proportions of the entire skeleton of the crocodile are adapted to the necessities of an amphibious animal, but minister to much more rapid and energetic movements in water than on land. The short limbs preclude the possibility of very quick course along shore; and the overlapping of the ribs of the neck, whilst enabling the head the better to cleave the water during the acts of diving or swimming, makes the bending of that part from side to side an act of difficulty and time; this, it is said, may avail any one pursued by a crocodile on dry land to escape by turning out of the straight course. But the crocodile usually seizes his prey by stratagem or concealment when in or close to the water; and it is there that he shows himself master of his position, and chiefly by the powerful strokes of his long, large, vertically-flattened tail.

OSTEOLOGY OF CHELONIAN REPTILES—TORTOISES AND TURTLES.

Those animals to which, in the manifold modifications of the organic framework, a portable dwelling or place of refuge has been given, in compensation for inferior powers of lomocotion or other means of escape or defence, have always attracted especial attention; and of them the most remarkable, both for the complex construction of their abode as well as for their comparatively high organization, are the reptiles of the chelonian order. The expanded thoracic-abdominal case, into which, in most chelonians, the head, the tail, and the four extremities can be withdrawn, and in some of the species be there shut up by movable doors closely fitting both the anterior and posterior apertures, as *e. g.* in the box-tortoises (*cinosternon*, *cistudo*), has been the subject of many and excellent investigations; and not the least interesting result has been the discovery that this seemingly special and anomalous superaddition to the ordinary vertebrate structure is due, in a great degree, to the modification of form and size, and, in a less degree, to a change of relative position, of ordinary elements of the vertebrate skeleton.

The natural dwelling-chamber of the chelonia consists chiefly, and in the marine species (*chelone*) and mud-turtles (*trionyx*) solely, of the floor and the roof; side-walls of variable extent are added in the fresh-water species (*emydians*) and land-tortoises (*testudinians*). The whole consists chiefly of osseous "plates" with superincumbent horny plates or "scutes," except in the soft or mud-tortoises (*trionyx* and *sphargis*), in which these latter are wanting.

Fig. 20 shows the manner in which the head and tail

can be retracted within the thoracic-abdominal box; the four limbs are figured as extended in the act of walking, to show their structure. The only movable vertebræ are those of the neck and tail, and the former enjoy a great degree of flexibility. The vertebræ answering to the dorsal, lumbar, and sacral series are firmly fixed together; but the dorsal ones, 1 to 8, are chiefly concerned in the formation of the osseous dwelling-chamber. The composition of this will be first described as it exists in the turtle (*chelone*), the species called "loggerhead" being here selected for its illustration.

In the marine species of the chelonian order, of which

Fig. 20.

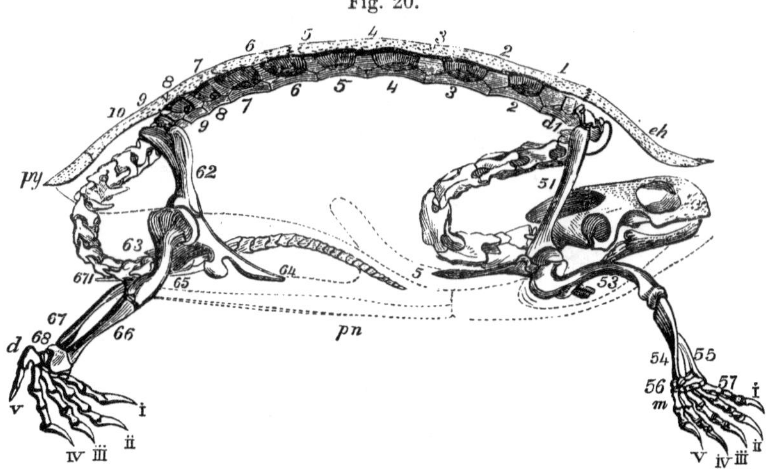

SKELETON OF THE EUROPEAN TORTOISE.

this may be regarded as the type, the ossification of the carapace and plastron is less extensive, and the whole skeleton is lighter than in the box-tortoise (Fig. 20), or any of those species that live on dry land. The head is proportionally larger—a character common to aquatic animals; and, being incapable of retraction within the carapace, ossification extends in the direction of the fascia covering the temporal muscles, and forms a second bony covering of the cranial cavity; this accessory defence is not due to the intercalation of any new bones, but to exogenous growths from the frontals, 11, postfrontals, 12, parietals, 7, and mastoids, 8.

The carapace (Fig. 21) is composed of a series of median and symmetrical pieces *ch*, *s* 1 to *s* 11, and of two series of unsymmetrical pieces on each side. The median pieces have been regarded as lateral expansions of the summits of the neural spines; the medio-lateral pieces as similar developments of the ribs; and the marginal pieces as the homologues of the sternal ribs. But the development of the carapace shows that ossification begins independ-

Fig. 21.

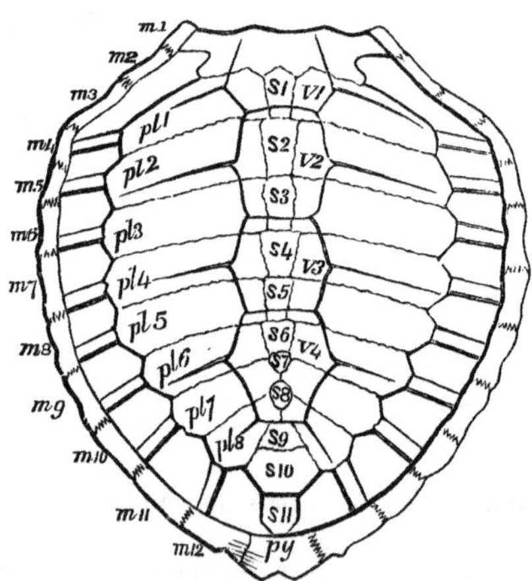

CARAPACE OF TURTLE (*Chelone imbricata*).

ently in a fibro-cartilaginous matrix of the corium in the first, *ch*, and some of the last, *s* 9 to *s* 11, median plates, and extends from the summits of the neural spines into only eight of the intervening plates, *s* 1 to *s* 8; ossification also extends into the contiguous lateral plates, *pl* 1 to *pl* 8, in some chelonia, not from the corresponding part of the subjacent ribs, but from points alternately nearer and further from their heads, showing that such extension of ossification into the corium is not a development of the tubercle of the rib, as has been supposed. Ossification commences independently in the corium in all the marginal plates, *m* 1 to *py*, which never coalesce with the bones uniting the sternum with the vertebral ribs, and which are often more numerous, and sometimes less numerous than those ribs, and in a few species are wanting. Whence it is to be inferred that the expanded bones of the carapace, which are supported and impressed by the thick epidermal scutes called "tortoise-shell," are dermal ossifications, homologous with those which support the nuchal and dorsal epidermal scutes in the crocodile. Most of the pieces of the carapace being directly continuous or connate with the obvious elements of the vertebræ, which have been supposed exclusively to form them by their unusual expansion, the median ones, *s* 1 to *s* 11, have been called "neural plates," and the mediolateral pieces, *pl* 1 to *pl* 8, "costal-plates;" but the external lateral pieces, *m* 1 to *m* 12, have retained the name of "marginal plates." The first or anterior of the median plates (*ch*, "nuchal plate") is remarkable for its great breadth in the turtles, and usually sends down a ridge from the middle line of its under surface, which articulates more or less directly with the summit of the neural arch of the first dorsal vertebra; the second neural plate is much narrower, and is connate with the summit of the neural spine of the second dorsal vertebra; the seven succeeding neural plates have the same relations with the succeeding neural spines; the rest are independent dermal bones. The costal plates of the carapace are superadditions to eight pairs of the pleurapophyses or vertebral portions of the second to the ninth ribs inclusive. The slender or proper portions of these ribs project freely for some distance beyond the connate dermal portions, along the under surface of which the rib may be traced, of its ordinary breadth to near the head, which liberates itself from the costal plate to articulate to the interspace of the two contiguous vertebræ, to the posterior of which such rib properly belongs.

The plastron, or floor of the bony house, consists in the genus *Chelone*, as in the rest of the order, of nine

Fig. 22.

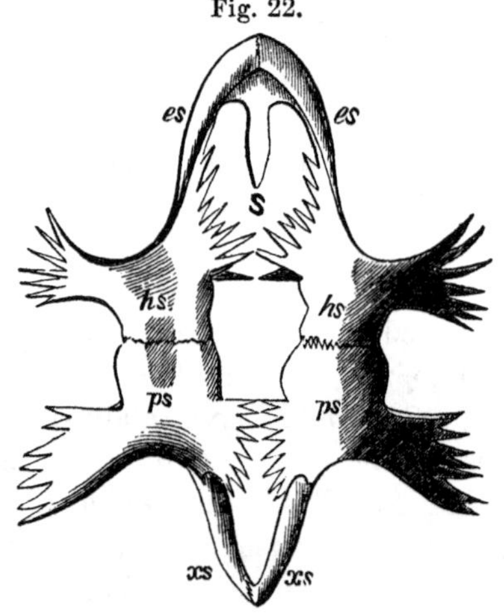

PLASTRON OF CHELONE CAOUANNA.

pieces—one median and symmetrical, and the rest in pairs. With regard to the homology of these bones, three explanations may be given: one in conformity with the structure of the thoracic-abdominal cage in the crocodile; the other based upon the analogy of that part in the bird; and the third agreeably with the phenomena of development. According to the first, the median piece of the plastron, called "entosternal," S, answers to the sternum of the crocodile, or "sternum proper," and the four pairs of plastron-pieces, *es*, *hs*, *ps*, *xs*, answer to the "hæmapophyses" forming the so-called sternal and abdominal ribs of the crocodile. Most comparative anatomists have, however, adopted the views of Geoffroy St. Hilaire, who was guided in his determination of the pieces of the plastron by the analogy of the skeleton of the bird; according to which all the parts of the plastron are referred to a complex and greatly developed sternum, and the marginal plates are viewed as sternal ribs (hæmapophyses). The third ground of determination refers the parts of the plastron, like those of the carapace, to a combination of parts of the endoskeleton with those of the exoskeleton.

In Fig. 21, the marginal plates *m* 1 to *m* 12, are twenty-four in number, or twenty-six if the first (nuchal, *ch*) and last (pygal, *py*) vertebral plates be included. Omitting these in the enumeration, three marginal pieces intervene on each side at the angles between the first median plate and the point of the first costal plate formed by the end of the second dorsal rib, which point enters a depression in the fourth marginal piece, *m* 4; the fifth, sixth, seventh, eighth, ninth, and tenth marginal plates are similarly articulated by gomphosis to the six succeeding ribs; the eleventh marginal plate has no corresponding rib; the twelfth is articulated with the point of the ninth dorsal rib supporting the eighth costal plate.

The want of concordance with the vertebral ribs, or "pleurapophyses," arising from the increased number of the marginal pieces, favors the idea of their being dermal ossifications, such peripheral elements being more subject to vegetative division and multiplication than the hæmapophyses: the absence of the marginal pieces in the trionyx gives additional support to the same view. The median piece, S, is here regarded as a hæmal spine: it is called "entosternum." The parial pieces of the plas-

Fig. 23.

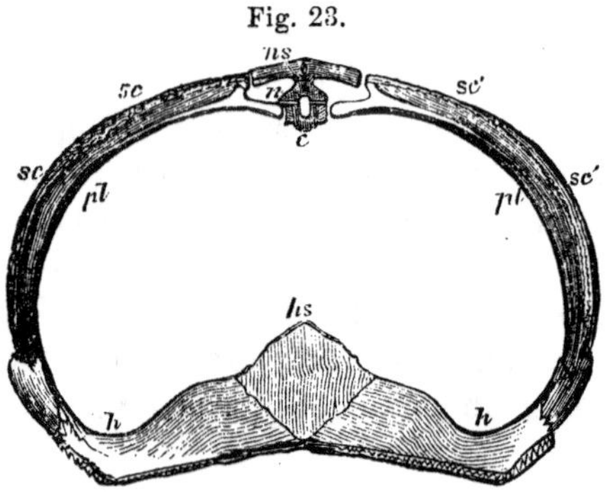

SEGMENT OF CARAPACE AND PLASTRON.

tron are the "hæmapophyses" connate with expanded dermal ossifications, and have received the following special names: *es*, "episternal;" *hs*, "hyosternal;" *ps*, "hyposternal;" *xs*, "xiphisternal."

In some extinct chelonia, the number of these lateral elements of the plastron is increased by an intercalated pair which I have called "mesosternals." In the figure of the segment, as modified to form the carapace and plastron (Cut 23), the nature of the bones is indicated by the letters according to the explanation given of the archetype vertebræ (Fig. 5, p. 29), the dermal superadditions being marked *sc*.

In the figure of the skeleton of the box-tortoise (Fig. 20) a section of the carapace and plastron has been removed from the right side to expose the dorsal and sacral vertebræ, and the disposition of the scapular and pelvic arches. The eight cervical vertebræ are free, movable, and ribless; the fourth of these vertebræ has a much elongated centrum, which is convex at both ends; the eighth is short and broad, with the anterior surface of the body divided into two transversely elongated convexities, and the posterior part of the body forming a single convex surface divided into two lateral facets; the under part of the centrum is carinate. The neural arch, which is anchylosed to this centrum, is short, broad, obtuse, and overarched by the broad expanded nuchal plate, *ch*. The first dorsal vertebra, *d* 1, is also short and broad, with two short and thick pleurapophyses, articulated by one end to the expanded anterior part of the centrum, and united by suture at the other end to the succeeding pair of ribs. The head of each rib of the second pair is supported upon a strong trihedral neck, and articulated to the interspace of the first and second dorsal vertebræ: it is connate, at the part corresponding to the tubercle, with the first broad costal plate, which articulates by suture to the lateral margin of the first neural plate, and to portions of the nuchal and third neural plates: the connate rib, which is almost lost in the substance of the costal plate, is continued with it to the anterior and outer part of the carapace, where it resumes its subcylindrical form, and articulates with the second and third marginal pieces of the carapace. The neural arch of the second dorsal vertebra is shifted forwards to the interspace between its own centrum and that of the first dorsal vertebra. A similar disposition of the neural arch and spine, and of the ribs, prevails in the third to the ninth dorsal vertebræ inclusive. The corresponding seven neural plates are connate with the spines of those vertebræ, and form the major part of the median pieces of the carapace; the corresponding costal plates, anchyclosed to the ribs, form the mediolateral pieces; the ninth, tenth, and pygal plates, with the marginal plates of the carapace, do not coalesce with any parts of the endoskeleton. The bony floor of the great abdominal box, or "plastron," is formed by the hæmapophyses and sternum connate with dermal osseous plates, forming, as in the turtle, nine pieces, one median and symmetrical, answering to the proper sternum, and eight in pairs: but they are more ossified, and the hyo and hyposternals unite suturally with the fourth, fifth, and sixth marginal plates, forming the side-walls of the bony chamber. The junction between the hyo- and hyposternals admits of some yielding movement. The iliac bones, 62, abut against the pleurapophyses of the tenth, eleventh, and twelfth vertebræ, counting from the first dorsal vertebra. These three vertebræ form the sacrum: their pleurapophyses are unanchylosed, converge, and unite at their distal extremities to form the articular surface for the ilium. Beyond these the vertebræ, thirty-five in number, are free, with short, straight, and thick pleurapophyses, articulated to the sides of the anterior expanded portions of the centrums. They diminish to mere tubercles in the first caudal vertebra, and disappear in the remainder. The neural arches of the caudal vertebræ are flat above, and without spines. The strong columnar scapula, 51, is attached by ligament to the first costal plate, and, retaining its primitive rib-like form, it descends almost vertically to the shoulder-joint, of which it forms, in common with the coracoid, 52, the glenoid cavity. A strong subcylindrical process or continuation of the scapula, representing the acromion, bends inwards to meet its fellow at the middle line. The coracoid continues distinct from the scapula, expands, and becomes flattened at its median extremity, which does not meet its fellow or articulate with the sternum. The iliac bones, 62, are vertical and columnar, like the scapula, but are shorter and more compressed: they articulate, but do not coalesce, with the pubis, 64, and ischium, 63. The acetabulum is formed by contiguous parts of all the three bones. The pubis arches inwards, and expands to join its fellow at the median symphysis and the ischium posteriorly. It sends outwards and downwards a long, thick, obtuse process from its anterior margin. The ischia, in like manner, expand where they unite together to prolong the symphysis backwards.

In the skull, the parietal crista is continued into the occipital one without being extended over the temporal fossæ, as in the turtle; the fascia covering the muscular masses in these fossæ undergoing no ossification. The bony hoop for the membrana tympani is incomplete behind, and the columelliform stapes passes through a notch instead of a foramen to attain the tympanic membrane. The mastoid is excavated to form a tympanic air-cell. In the Australian long-necked terrapene (*hydraspis longicollis*) the head is much depressed, the mastoids are excavated by large tympanic cells, and prolonged backwards: the frontal is produced forwards as far as the anterior nostril, where it terminates in a point between the two nasals, which are here distinct from the prefrontals. The margins of the upper and lower jaws are trenchant: the hypapophysis of the atlas has the form of a diminutive wedge-bone, forming as usual the lower part of the articular cup for the occipital condyle: the rest of the body of the atlas, or "odontoid," has coalesced with its proper neural arch, which develops two transverse and two long posterior oblique processes, as in the chelys.

In the true or land tortoises the temporal depressions are exposed, as in the box-tortoises and fresh-water terrapenes: the head is proportionally small, and can be withdrawn beneath the protective roof of the carapace. The skull is rounder and less depressed than in the terrapenes: the frontals enter into the formation of the orbital border. The tympanic hoop is notched behind, but the columelliform stapes passes through a small foramen. The palatine processes of the maxillaries are on a plane much below that of the continuation of the basis cranii, formed by the vomer and palatines. In most of the chelonia, the nasal bone is connate with the prefrontal; and, in all, the tympanic pedicle is firmly wedged between the broad appendage of the maxillary arch, formed by the malar, 26, and squamosal, 27, in front, and the mastoid, 8, behind. The broad-headed terrapene (*podocnemys expansa*) differs from other fresh-water tortoises, and approaches the marine tortoises (turtles), by the vaulted bony roof arching over the temporal depressions. This roof is chiefly formed by the parietals, but differs from that in the turtles in being completed laterally by a larger proportion of the squamosal than of the postfrontal, which does not exceed its relative size in other terrapenes. The present species further differs from the marine turtles in the nonossification of the vomer, and the consequent absence of a septum in the posterior nostrils; in the greater breadth of the pterygoids, which send out a compressed rounded process into the temporal depressions: the orbits also are much smaller, and are bounded behind by orbital processes of the postfrontal and malar bones: the mastoids and paroccipitals are more produced backwards, and the entire skull is more depressed than in the turtles.

The ordinary position of the scapular extremity is a state of extreme pronation, as shown in Fig. 20, with the olecranon, or top of 54, thrown forwards and outwards, and the radial side of the hand, or thumb, *i*, directed to the ground. The humerus, 53, is strongly bent in a sigmoid form, with the anconal surface convex and directed upwards and outwards: the two tuberosities at the proximal end are much developed and bent towards the palmar aspect, bounding a deep and wide groove: that which answers to the external tuberosity is the smallest, and by the rotation of the humerus it becomes the most internal in position. The proximal row of the carpus consists of four bones—viz: a large scaphoides, a small lunare, wedged into the interspace of the radius and ulna, a large cuneiforme, and a small pisiforme. The second row consists of five distinct bones, corresponding with the five digits; those supporting the fourth and fifth answering to the os unciforme, the remaining three to the trapezium, trapezoides, and magnum. The first and fifth of the digits have each one metacarpal and two phalanges; the rest, *ii*, *iii*, *iv*, have each a metacarpal and three phalanges. A sesamoid bone is placed beneath the metacarpo-phalangeal joint of the three middle digits.

In the pelvic extremity, the femur, 65, is sigmoidally bent, but in a less degree than the humerus, and is a shorter bone. The patella is ligamentous: the synovial joint between it and the femur is distinct from the proper capsule of the knee-joint; the fibula, 66, is longer and more slender than the tibia, 66; a small "fabella" is articulated to its upper end. The proximal row of the tarsus consists of two bones, astragalus and calcaneum, which sometimes become confluent. The distal row consists of five bones, four of which support the four normal toes, and the fifth, a rudiment of the fifth toe without a claw: the fourth and fifth of the second row of tarsals answer to the os cuboides of higher animals; the other three bones to the three ossa cuneiformia. The astragalar part of the single proximal bone would seem to include the scaphoid as well as the calcaneum.

In the marine chelonia, the digits of both limbs are elongated, flattened, and united by a web; the hands and feet having the form of fins.

In all the chelonia, the long bones of the limbs are solid, without medullary cavities.

THE SKELETON OF BIRDS.

From the massive frame of the cold-blooded, heavy, and proverbially slow tortoise, to the light, hot-blooded, flying bird, the transition seems to be abrupt, and the discrepancy between creatures so differently endowed extreme; nevertheless, at the confines of the feathered class, we find some aquatic species, such as the penguin, incapable of flight, having the wings modified to act as fins, and much resembling those of the turtle; with the bones solid, and the feathers resembling scales. All birds, like tortoises, lay eggs, are devoid of teeth, and have their jaws sheathed with horn, and forming a bill or beak. Most birds, however, enjoy the faculty of flight.

If the student of comparative osteology will procure the skull of a rook, a hawk, a swan, or a sea-gull, and vertically bisect it, he will have a ready instance illustrative of some of the characteristics of the osteology of the feathered class. Such a section will show the ivory-like whiteness and compactness of the osseous tissue, and the loose, open, cancellous structure of the bones. He will see that air is admitted into these cancelli partly from the nasal passages, and partly from the tympanic cavity which receives it from the Eustachian tube; from the latter source, the proper bones of the cranium receive their air. Some of the characteristic features in the composition of the skull of birds may also be noticed: as, for example, the obliteration of all the ordinary sutures of the cranium, except those which unite the tympanic bone, 28, to the mastoid, 8; and that which unites the pterygoid, 23, to the basisphenoid, 5; which sutures are speedily obliterated in the human subject. The premaxillary is confluent with the nasal and with the maxillary; the nasal being confluent with the frontal and the maxillary with the jugal. The jugal and squamosal are also confluent, and form a long zygomatic style in all birds, connected at the hinder extremity by a movable glenoid joint to the outer and lower part of the tympanic. The pterygoid articulates, in like manner, with the inner and lower part of the tympanic, the movements of which are thus communicated to the upper mandible, so far as the junction of the nasal with the frontal admits of such independent motion. The upper jaw, or mandible, which includes the vomer and nasals with the maxillary arch and appendages, is movable in a bird through the junction of the nasals and nasal branch of the premaxillary with the frontal, by means of a movable articulation, or by elastic plates.

If the student will next separate one of the vertebræ of the trunk from the rest, and cut out that portion of the long and broad breast-bone to which its pair of ribs are attached, he will have a segment of the skeleton, answering to that figured in Fig. 5, p. 28.

The cut surfaces will demonstrate the light cellulosity of the divided bones. The following letters indicate the elements of such modified vertebræ of the thorax: *y*, centrum, with its hyapophysis; *p*, parapophysis; *d*, diapophysis; *n*, neural arch and rudimental spine; *pl*, pleurapophysis; *h*, hæmapophysis; *hs*, hæmal spine. The tendency of individual elements and bones to coalesce in birds has already been illustrated in the cranium; it is shown, in most birds of flight, not only by the confluence of the centrum with the neural arch, but by that of several consecutive centrums and arches into a single bone, in the ample chest. In like manner the hæmal spines, which continue distinct in many vertebrata, have here coalesced into a single bone, which articulates on each side with the hæmapophyses of several vertebræ. These coalesced spines are also much developed in breadth, and send down, from the middle of their under surface, a longitudinal crest or keel. This modification relates to the extension of the surface for the origin of the great muscles of flight, and renders the "sternum," as the coalesced series of hæmal spines is called, one of the most characteristic parts of the skeleton of the bird. Ossification extends from the neural arches into the tendons of the vertebral muscles, and such bone-tendons, both here and in other parts of the body, as the legs, are also characteristic of birds. The scapula (Fig. 24), 51, is long and slender, as in the chelonia, but is more compressed and sabre-shaped. The coracoid, 52, as a general rule, is a distinct bone, movably articulated to the scapula at one end and to the sternum at the other. Its broad sternal end here articulates by a kind of gomphosis with a deep

Fig. 24.

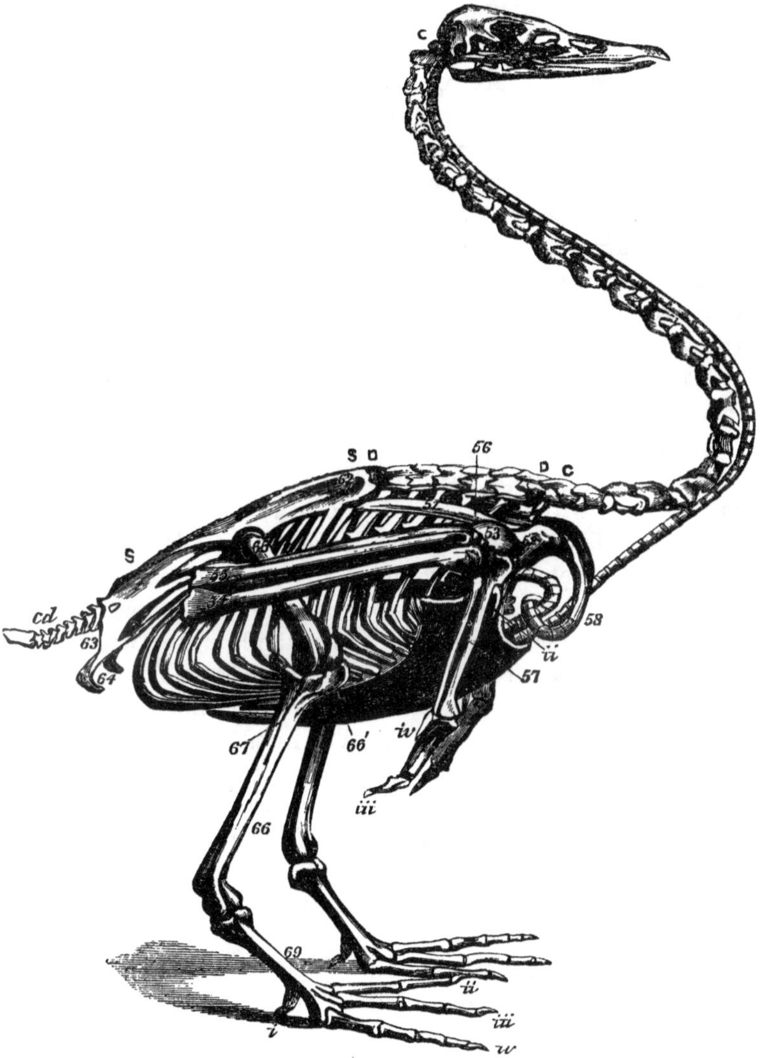

SKELETON OF THE SWAN (*Cygnus ferus*).

groove on the fore part of the sternum. The clavicle (*ib.*), 58, articulates with the coracoid above, but is confluent with its fellow and with the keel of the sternum below. The iliac bones, 62, are remarkable for their length, and for the number of the vertebræ, or the great extent of the confluent spinal column, to which they are anchylosed. They reach in the swan, and in most other birds, from the tail forwards to the vertebræ with movable ribs. Thus the artificial characters of a "lumbar vertebra" are wanting. The pubis and ischium on each side have coalesced with the ilium to form the lower boundary of the widely-perforated acetabulum. The pubis is long and slender, joins the ischium of its own side near its lower extremity, but does not join its fellow; thus the foramen ovale is defined, but there is no symphysis pubis: the absence of this symphysis facilitates the expulsion of the large ovum with its unyielding calcareous shell. The ischium coalesces posteriorly with the ilium, and converts the ischiadic notch into a foramen. The caudal vertebræ, *Cd*, are few in number, with broad transverse processes formed by confluent pleurapophyses, the limits of which may still be traced. A hæmapophysis is articulated to the lower interspace, between the fourth and fifth caudal, and is anchylosed to the sixth. The humerus of some of the larger birds of flight—*e. g.* the pelican or adjutant crane—is remarkable for its lightness, as compared with its bulk and seeming solidity; it is, in fact, a mere shell of compact osseous tissue. The orifice admitting air to its large cavity is beneath the great tuberosity at the proximal end.

The keel is excavated, not only for the reception of an air-cell, but likewise for a fold of the windpipe, which fold expands with age, and lies horizontally in the substance of the back part of the sternum. Small pneumatic foramina are situated at the anterior and inner surface of the bone, and perforate the articular surfaces for the sternal ribs.

In the skeleton of the wild swan (*Cygnus ferus*) (Fig. 24), here selected as an illustration of the ornithic modification of the vertebrate type, there are not fewer than twenty-eight vertebræ, C S D, between the skull and the sacrum, the last six of which, D D, support movable ribs; of these, the first and second pairs are free; the next four are articulated to the sternum by bony hæmapophyses; the last five pairs of ribs are attached to the sacrum, and also to the sternum; but the tenth, or last rib on the left side, is very rudimentary, being only about one inch in length. There are eight caudal vertebræ, *Cd.* The trachea, or windpipe, penetrates the sternum, and bends and winds in the interior of the bone before returning to enter the chest. The apex of the furculum, 58, bends upwards, and forms a hoop over the windpipe as it enters into the keel of the breast-bone. The furculum, sometimes called "merrythought," consists of the two clavicles confluent at their lower free ends. If a portion of the one side of the sternum be removed, the tortuous trachea which it incloses will be exposed. To the great length and peculiar course of the windpipe in this species is to be attributed its remarkably loud and harsh voice; whence the name hooper, or whistling swan, has been derived; and is applied in contradistinction to the domestic or mute swan, in which, as in most other birds, the trachea proceeds at once to the lungs, without entering the sternum. In the female of the wild species, the course of the trachea is much more limited than in the male, seldom penetrating the sternum to a greater extent than from three to four inches.

The breadth of the sternum, and the strong ridge or keel that descends from the midline of its under surface, relate to the increased extent of surface required for the attachment of the "pectoral" muscles, which are the active organs of flight. In the land-birds devoid of the power of flight, such as the ostrich and apteryx, the keel is wanting and the sternum is short. Its various proportions, processes, notches, and perforations render it a very characteristic bone in birds.

In no order, founded upon modifications of the feet, is the sternum more diversified in character than in the palmipedes or web-footed order; for in none are the powers of flight enjoyed in such different degrees, or exercised in such various ways, from the frigate-bird down to the penguins, where the power of flight is abrogated, and the rudimental wings used as fins.

In the goose and duck tribes, as well as the swans (*anseres*, Linn.), the sternum is long and broad, and presents two moderately wide and deep hind notches; the costal processes are usually subquadrate; the coracoid grooves are continued into one another at the median line; the costal tract forms about half of the lateral margin in the ducks and geese, and two-thirds or more in the swans; the interpectoral ridge extends from the prominent part of the coracoid margin backwards, nearly parallel to the lateral margin, to the inner side of the lateral grooves; the back part of the sternum between the grooves is quadrate, with the angles slightly produced in most; there is a short manubrial process below the coracoid groove. The form of the sternum, its long keel, and the backward production of the long and slender ribs, give a boatlike figure to the trunk of these swimming-birds, which is well adapted to their favorite medium and mode of locomotion. The bones of the wing or anterior extremity do not present that extraordinary development which might be expected from the powers of the member of which they form the basis. The great expanse of the wing is gained at the expense of the epidermoid system (quills and feathers, like hairs and scales, are thickened epiderm), and is not exclusively produced by folds of the skin requiring elongated bones to support them, as in the flying-fish, flying-lizards, and bats. The wing-bones of birds are, however, both in their forms and modes of articulation, highly characteristic of the powers and applications of the muscular apparatus requisite for the due actions of flight. The bones of the shoulder consist on each side of a scapula, 51, a coracoid, 52, and a clavicle, 58, the clavicles being, as a general rule in birds, confluent at their median ends, and so forming a single bone called "furculum," or "os furcatorium;" this further modification of the hæmal arch in birds, repeating that of the pubis and lower jaw in some other animals, having occasioned an additional specific term in ornithotomy. The scapula, 51, is a long, narrow, flat sabre-shaped plate, expanded at the humeral end, where it forms externally part of the joint for the arm-bone called "glenoid cavity," and extended backwards nearly parallel with the vertebræ, as far as the ilium, 62, in the swan, and reaching to the last rib in the swift; but it is much shorter in the birds incapable of flight. The coracoid is the strongest of the bones of the scapular arch: it forms the anterior half of the glenoid cavity, extends above this part to abut upon the furculum, and is continued downwards below the joint, expanding, to be fixed in the transverse groove at the fore part of the sternum; it thus forms the chief support of the wing, and the main point of resistance during its downward stroke. In the hawks and other birds of prey, and in the crows and most passerine birds, a small bone (os humero-capsulare) extends between the scapula and coracoid along the upper part of the glenoid cavity; this is absent in the swan and other swimmers, as well as in the gallinaceous and wading birds. The humerus, 53, is usually a long and slender bone, but is not always developed in length in proportion to the powers of flight; for, although it is shortest in the struthious birds and penguins, it is also very short, but much thicker and stronger in the swift and humming birds. The head of the humerus is transversely oblong and convex; it is further enlarged by two lateral crests; of these the superior is the longest, and is bent outwards; the inferior is thickened and incurved, and beneath it is situated the orifice by which the air penetrates the cavity of the bone. The articular surface at the opposite or "distal" end is divided into two parts, one internal, for the ulna, of a hemispheric form, the other also convex, but more elongated and oblique, extending some way upon the anterior surface of the humerus. The extremity of a long bone of a limb which is next the trunk is called the "proximal" one; the extremity furthest from the trunk, the "distal" one; they are not always "upper" and "lower." The ulna, 55, glides upon the inner hemispheric tubercle, upon the trochlear canal, and on the back part of the outer convexity. A ligament, extending from the outer part of the head of the radius to the outer part of the olecranon, above the posterior margin of the outer division of the articular surface of the ulna, plays upon the back part of the radial convexity of the humerus and completes the cavity receiving it. The ulna is always stronger than the radius; but both are long, slender, and nearly straight bones, so articulated together as to admit of scarcely any rotation which adds to the resisting power of the wing in the action of flight. The upper part of the ulna, or "olecranon," is short. In the tendon attached to it, a separate ossicle is developed in the swift, and two such bones in the penguin. The ulna is often impressed by the insertions of the great quill-feathers of the wing.

The bones of the hand are very long and narrow, with the exception of the two distinct or unanchylosed carpal bones; these are so wedged in between the antibrachium, 54, 55, and the metacarpus, 57, as to limit the motions of the hand to abduction and adduction, or those necessary for folding up and spreading out the wing. The hand is thus fixed in a state of pronation; all power of flexion, extension, and rotation is removed from the wrist-joint; so that the wing strikes firmly, and with the full force of the depressed muscles, upon the resisting air. The part of the hand numbered 57 in Fig. 24 includes the metacarpal bones of the digits answering to the second, third, and fourth of the pentadactyle members, which are confluent at their proximal ends with each other, and with the "os magnum," one of the carpal bones, now forming the convex base of the middle metacarpal. This metacarpal, and that answering to the "fourth" digit, are of equal length, and are also confluent at their distal ends; but the middle or "third" metacarpal is much the strongest. That answering to the "second" digit, *ii*, is very short, and like a mere process from the third; it supports two short phalanges in the swan. The third metacarpal supports three phalanges, *iii*, the fourth a single phalanx, *iv*. All these are wrapped up in a sheath of integument, and are strongly bound together; so that the wing loses nothing of its power, whilst so much of the typical structure of the member is retained, that every bone can be referred to its corresponding bone in the most completely developed hand.

In ornithology, the large quill-feathers that are attached to the ulnar side of the hand are termed "primariæ," or primary feathers; those that are attached to the forearm are the "secundariæ," or secondaries, and "tectrices," or wing-coverts; those which lie over the humerus are called "scapulariæ," or scapularies; and those which are attached to the short outer digit, *ii*, erroneously called the "thumb," are the "spuriæ," or bastard feathers. The bones of the leg do not present the same number of segments as those of the wing, that corresponding with the carpus being wholly blended with the one that succeeds.

The pelvic bones offer this contrast with those of the shoulder, that they are always anchylosed on either side into one piece, "os innominatum," and not at the median line, whilst this is the only place where the elements of the scapular apparatus are united by bone. In the young bird, the os innominatum is composed of three bones. The ilium, 62, is flattened, elongated, usually anchylosed to a very long sacrum; it forms the upper half of the joint for the thigh-bone, called "cotyloid cavity." The pubis, 64, is very long and slender; it does not meet its fellow at the middle line in any bird save the ostrich, but is directed backwards, with its free extremity bent downwards. The pelvis of the ostrich is so vast, that the pubic junction completing it does not impede the exit of the egg; in other birds the open pelvis facilitates the passage of that large and brittle generative product. The ischium, 63, is a simple elongated bone, extending from the cotyloid cavity backwards, parallel with the ilium; it sometimes coalesces, as in the swan, with both the ilium and pubis at its distal end.

The cotyloid cavity is incomplete behind, and is closed there by ligament. The femur, 65, is a short, cylindrical, almost straight bone; the head is a small hemisphere, presenting at its upper part a depression for the "round ligament." The single large "trochanter" generally rises above the articular eminence, and is continuous with the outer side of the shaft. The orifice for the admission of air is situated in the depression between the trochanter and head. The distal end presents two condyles, the inner one for the inner condyloid cavity of the tibia; the outer one for the outer cavity of the tibia and for the fibula; the outer condyle is produced into a semicircular ridge, which passes between the tibia and fibula; this ridge puts the outer elastic ligament on the stretch, when the fibula is passing over the condyle, and the fibula is pulled into a groove at the back of the condyle, with a jerk, when in extreme flexion; this spring-joint is well exemplified in both the swan and water-hen.

The proximal end of the tibia is divided into the two shallow condyloid cavities above noticed: two ridges are extended from its upper and anterior surface: the strongest of these is the "procnemial" ridge, and is slightly bent outwards; the shorter one on the outside of this is the "ectocnemial" ridge; they are usually united above by a transverse ridge, called "epicnemial" ridge; this is developed into a long process in the divers, grebes, and guillemots: a fibular ridge projects slightly from the upper third of the tibia for junction with the fibula. The distal end of the tibia forms a transverse pulley or trochlea, with the anterior borders produced. Above the fore part of the trochlea is a deep depression, and in many birds an osseous bridge extends across it.

The third segment of the leg, 69, is a compound bone, consisting originally of one proximal piece, short and broad, presenting two articular concavities to the two thick and round borders of the tibial trochlea, of three metatarsals which coalesce with each other and with the above tarsal piece, and of one or more bony processes which are ossified from the back part of the proximal piece, or from the proximal ends of the metatarsals, and which, from their relations to the extensor tendons, are called "calcaneal" processes. In most birds, a small rudimental metatarsal, supporting the innermost toe, or "hallux," *i*, is articulated by ligament with the innermost of the coalesced metatarsals, and is properly included in the same segment of the limb. The three principal metatarsals are interlocked together before they become anchylosed, the middle one being wedged into the back part of the interspace of the two lateral ones above, and into the fore part below, passing obliquely between them. The period at which these several constituents of the "tarso-metatarse" coalesce is shorter in the birds that can fly than in those that cannot; and the extent of the coalescence is least in the penguins, in which the true nature of the compound bone is best seen.

The modifications of the tarso-metatarse are chiefly manifested in its relative length and thickness, in the relative length of the three metatarsals, and in the number and complexity of the calcaneal processes.

The inner of the two cavities for the condyles at the proximal end of the bone is the "entocondyloid" cavity or surface, the outer one the "ectocondyloid" surface; they are separated by an "intercondyloid" tract, from the forepart of which there usually rises an intercondyloid tuberosity. The entocondyloid cavity is usually the largest and deepest: it is so in the raven, in which the base of the intercondyloid tubercle extends over the whole of the intercondyloid space. There are three calcaneal processes: one, called the "entocalcaneal," projects from below the entocondyloid cavity, and from the back part of the upper end of the entometatarse; a second, called the "mesocalcaneal," from the intercondyloid tract and the mesometatarse, and the third called "ectocalcaneal," from behind the ectocondyloid cavity and the ectometatarse. These three processes are united together by two transverse plates circumscribing four canals, two smaller canals being further carried between the ento and meso-calcaneal processes. The primitive interosseous spaces are indicated by two small foramina at the upper and back part of the shaft, which converge as they pass forward, and terminate by a single foramen at the fourth part of the anterior concavity. A similar minute canal is retained between the outer and middle metatarsals, near their distal ends; each metatarsal then becomes distinct, and develops a convex condyle for the proximal phalanx. The middle one is the largest, and extends a little lower than the other two: it is also impressed by a median groove; the more compressed lateral condyles are simply convex, and are of equal length. A rough surface, a little way above the inner condyle, indicates the place of attachment of the small metatarsal of the hallux.

In the swan and other anserine birds the calcaneal prominence presents four longitudinal ridges, divided by three open grooves, the innermost ridge being the largest; the shaft is subquadrate, with the angles rounded, and none of the surfaces are channelled. The inner condyle scarcely extends before the base of the middle one; the canal perforating the outer intercondyloid space is bounded below by two small bars passing from the middle to the outer condyle, and which bars define the groove for the abductor muscle of the outer toe.

The tarso-metatarse of the diver (*colymbus*) is remarkably modified by its extreme lateral compression. The ento and ecto-calcanea are prominent, oblong, subquadrate plates, inclining towards each other, but not quite circumscribing a wide intermediate space. The broad outer and inner surfaces of the shaft are nearly flat; the narrow fore and back surfaces are channelled; the anterior groove leads to the wide canal, perforating obliquely the shaft above the outer intercondyloid space, from which a narrower canal conducts to that interspace. The middle and outer trochleæ are nearly equally developed; the inner one stops short at the base of the middle one.

The number of toes varies in different birds; if the spur of the cock be regarded as a rudimental toe (which is not, however, my view of it), it may be held to have five toes, while in the ostrich the toes are reduced to two. Birds, moreover, are the only class of animals in which the toes, whatever be their number or relative size, always differ from one another in the number of their joints or phalanges, yet at the same time present a constancy in that variation.

The innermost or back toe, *i* (Fig. 24), answering, as I believe, to the "hallux," or innermost digit of the pentadactyle foot, has two phalanges; the second toe, *ii*, has three, the third toe, *iii*, four, and the fourth toe, *iv*, five phalanges. I believe the toe answering to the fifth in lizards and other pentadactyle animals to be wanting in the bird's foot; and the spur, sometimes single, sometimes double, as in *Pavo bicalcaratus*, to be a superadded weapon to the metatarse. As the toes in the tridactyle emeu, cassowary, and bustard, have respectively three phalanges, four phalanges, and five phalanges, we recognize them as answering to the second, third, and fourth in other birds; the toes in the didactyle ostrich have respectively four and five phalanges, and what is here truly suggestive, the outermost, which is much the smallest and shortest toe, has the greater number of joints, viz: five, thus retaining its ornithic type, as the fourth, or outermost, toe.

The entire form of the body, and consequently that of its bony framework, in a bird, has special reference to the power of flight. The trunk is an oval with the large end forwards. The vertebral column of this part is short and almost inflexible, so that the muscles act to great advantage; the spine of the neck being long and flexible, the centre of gravity is readily changed from above the feet—as when standing or walking—to between and beneath the wings during flight; when suspended in the air the bird's body naturally falls into that position, which throws the centre of gravity beneath the wings. The axis of motion being situated in a different place in the line of the body when walking from that which is used when flying, the discrepancy requires to be compensated by some means in all birds, in order to enable them to perform flight with ease. Raptorial birds take a horizontal position when suspended in the air, and the compensating power consists in their taking a more or less erect position when at rest. Another class, including the woodpeckers, wagtails, &c., take an oblique position in the air; with these the compensating power consists in their cleaving and passing through the air at an angle coincident with the position of the body, and performing flight by a series of curves or saltations. Natatorial birds sometimes need very extended flight; they take a very oblique position in the air, stretch out their legs behind and their neck in front; they have the ribs greatly lengthened, the integuments of the abdomen are long and flexible, which enables them greatly to enlarge the abdominal portion of their bodies by inflating it with air; this causes a decrease in the specific gravity of that part, and raises it to a horizontal position; the compensating power consists in the posterior half of the body becoming specifically lighter, while the specific gravity of the anterior half remains unaltered. When they alight they drop the legs, throw back the trunk by bending the knee-joint, and bring the head over the trunk by a graceful sigmoid curve of the long neck, as in Fig. 24. The act of swimming is rendered easy by the specific gravity of the body, by the boatlike shape of the trunk, and by the conversion of the hinder extremities into oars, in consequence of the membranes uniting the toes together. The effect of these web-feet in water is further assisted by the toes having their membranes lying close together when carried forwards; whilst, on the contrary, they are expanded in striking backwards. The oarlike action of the legs is still further favored by their backward position—an arrangement, however, which is unfavorable for walking.

Borelli was the first who, by comparison of the anatomical peculiarities of the human frame and the structure of birds, demonstrated, to a certain extent, the impossibility of the realization of the cherished project of flying by man. He arrived at this conclusion from a comparison of the form and strength of the muscles of the wings of birds with the corresponding muscles of the human body.

PRINCIPAL FORMS OF THE SKELETON IN THE CLASS MAMMALIA.

In the class Mammalia, which includes the hairy quadrupeds with the naked apodal whales and biped man, the form of the animal is modified for a great diversity of kinds and spheres of locomotion. Some live exclusively in the ocean, and cleave the liquid element under the form and with the locomotive powers of fishes; some frequent the fresh waters; some pass a subterraneous existence, and work their way through the solid earth; some mount aloft, to seek and seize their prey in the air; some pass their lives in trees; most, however, dwell on the earth, with various powers of walking, running, and leaping. Lastly, man is modified to sustain his frame erect on the hinder, now become in him the lower, limbs.

In the Mammalian class, accordingly, we find the limbs progressively endowed with more varied and complicated powers. They retain in the *Cetacea* (whale and porpoise tribe) their primitive form of flattened fins; in the *Ungulata* (hoofed beasts) one or more of the digits acquire the full complement of joints, but have the extremity enveloped in a dense hoof; in the *Unguiculata* (quadrupeds with claws), the limbs, with ampler proportions, have the digits liberated, and armed with claws confined to the upper surface, leaving the under surface of the toes free for the exercise of touch; in the mole, the hand is shortened, thickened, expanded, and converted into a sort of spade; in the bat, the fingers are lengthened, attenuated, and made outstretchers and supporters of a pair of wings; in the *Quadrumana* (ape and monkey tribes), certain digits are endowed with special offices, and by a particular position enabled to oppose the others, so as to seize, retain, and grasp. Lastly, in *Man*, the offices of support and locomotion are assigned to a single pair of members; the anterior, and now the upper, limbs being left free to execute the various purposes of the will, and terminated by a hand, which, in the matchless harmony and adjustment of its organization, is made the suitable instrument of a rational being.

In contemplating and comparing the skeletons of a series of mammals, the most striking modifications are observable in the structure and proportions of the limbs.

There are a few osteological characters in which all mammalia agree, and by which they differ from the lower vertebrata; and some have been supposed to be peculiar to them that are not so. The pair of occipital condyles, *e. g.*, developed from the exoccipitals, are a repetition of what we saw in the batrachia. The flat surfaces of the bodies of the trunk-vertebræ were a character of many extinct reptiles; but these surfaces in mammals are developed on separate epiphysial plates, which coalesce in the course of growth with the rest of the centrum. Movable ribs, projecting freely (pleurapophyses) in the cervical region, may be found in a few exceptional cases (sloths, monotremes); bony sternal ribs (hæmapophyses) exist in most *Edentata;* a coracoid extending, as in birds and lizards, from the scapula to the sternum, with an "epicoracoid," as in lizards, is present in the monotremes (platypus or duck-mole, and echidna or spiny ant-eater, of Australia); the cotyloid cavity may be perforated in the same low mammals as in birds; the digits may have the phalanges in varying number in the same hand, and exceeding three in the same finger, *e. g.*, in the whale tribe. But the following osteological characters are both common and peculiar to the mammalia. The squamosal, 27, or second bone of the bar continued backwards from the maxillary arch, is not only expanded as in the chelonia, but develops the articular surface for the mandible, and this surface is either concave at some part or is flat. Each half or ramus of the mandible is ossified from a single centre, and consists of one piece; and the condyle is either convex or is flat, never concave. The presphenoid (centrum of the parietal vertebra) is developed distinctly from the basisphenoid; it may become confluent, but is not connate, therewith.

One known mammal (the three-toed sloth) has more, and one (the manatee, or sea-cow) has less than seven vertebræ of the neck. In the rest of the class these vertebræ, which have the pleurapophyses short and usually anchylosed, are seven in number.

SKELETON IN THE CETACEA, OR WHALE TRIBE.

In the skeleton of the whale (Fig. 25), which to outward appearance seems to have as little neck as a fish, there are as many cervical vertebræ as in the long-necked giraffe: this is a very striking instance of adherence to type within the limits of a class: the adaptation to form and function is effected by a change of proportion in the bones; the cervical vertebræ in the whale are flattened from before backwards into broad thin plates; in the giraffe (Fig. 30) they are produced into long subcylindrical bones. In the whales, the movements of these vertebræ upon one another are abrogated; and in the grampus and porpoise, the seven vertebræ are blended together into a single bone; they thus give a firm and unyielding support to the large head, which has to overcome the resistance of the water when the rapid swimmer is cleaving its course through that element. The dorsal

Fig. 25.

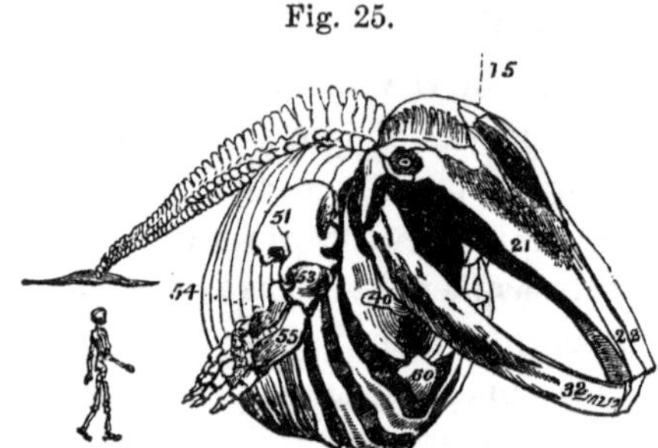

FORESHORTENED VIEW OF THE SKELETON OF A WHALE (*Balænoptera boops*), SHOWING ITS RELATIVE SIZE TO MAN.

vertebræ are characterized in all mammalia by the sudden increase in the length and size of the ribs, which, in a certain number of these vertebræ, including the first, are joined to a breast-bone by a commonly cartilaginous, rarely osseous, part. The first rib is remarkable for its great breadth in the whale; this and a few following ribs are joined to a short and broad and often perforated sternum (Fig. 25), No. 60; the remaining ribs are free, or, as they would be called in Human Anatomy, "false." They are articulated to the ends of diapophyses, which progressively increase in length to the last of the dorsal series. Then follow vertebræ without ribs, answering to those called "lumbar." The whole hinder part of the trunk of whales being needed to effect the strokes by which they are propelled, its vertebræ are as free from anchylosis as in fishes; there is consequently no "sacrum," and the caudal vertebræ are counted from the first of those that have "chevron bones" articulated to their under part. This special name is given to the vertebral elements called "hæmapomophyses" (see Fig. 26, *h*), which are articulated in cetacea as in crocodilia, directly to the under surface of the centrum, and, coalescing at their opposite ends, develop thence a "hæmal spine," and form a "hæmal" canal analogous to, but not homologous with, that in fishes (compare No. V, *h*, with No. I, *p*, in Cut 10, p. 182). The caudal vertebræ of whales further differ from those in fishes in retaining the transverse processes, and in becoming flattened from above downwards, without coalescing. These modifications relate to the support of a caudal fin, which is extended horizontally instead of vertically.

Whales and porpoises progress by bounding movements or undulations in a vertical plane, and their necessity of coming to the surface to inhale the air directly, as warm-blooded mammals, calls for a modification in the form of the main swimming instrument, such as may best adapt it to effect an easy and rapid ascent of the head.

The course of the whale is stopped and modified by the action of the pectoral limbs, which are the same parts as those in fishes, but constructed more after the higher vertebrate type. The digital rays do not exceed five in number; but they consist of many flattened phalanges, and are enveloped in a common sheath of integument. A radius, 55, and an ulna, 54 (Fig. 25), support the carpal series; but, instead of being directly articulated to the scapular arch, they are suspended to a humerus, 53: this is a short, thick bone, with a rounded head. The scapula, 51, is detached from the occiput, has a short, stunted, coracoid anchylosed to it, and is thus freely suspended in the flesh; it develops an acromial process: the ulna, 54, is produced upwards into an olecranon. With all those marks, however, of adhesion to the mammalian type of forearm, the outward aspect of the limb is as simple as is that of the fish's fin; it moves, as by one joint, upon the trunk, and is restricted to the functions of a pectoral fin.

In the huge skull of the whale, the broad vertical occiput may be noticed, by which the head is connected, through the medium of a short, consolidated neck, with the trunk; the whole cranium seems to have been compressed above, from before backwards, so that the small nasal bones, 15, articulating with the short and very broad frontals, form the highest part of the skull. The long maxillaries, 21, and premaxillaries, 22, extend backwards and upwards, to articulate with the nasals, and complete with them the bony entry to the air-passages, situated so favorably at the summit of the cranium. The nostrils, formed by the soft parts guarding that entry, are called "blow-holes;" they are double in the whales—single in the smaller cetacea. In the whales, the "baleen," or "whalebone" plates are attached to the palatal surface of the maxillary and premaxillary bones; the expanded toothless mandible supports an enormous under lip, which covers the whalebone plates when the mouth is shut. The skeleton of the great finner whale (*Balænoptera boops*), from which the foreshortened view (Cut 25) is taken, was ninety-six feet in length; the relative dimensions of man is given by the outlines of the skeleton at its side. No known extinct animal of any class equalled this living Leviathan in bulk.

There are a few whalelike mammals, equally devoid of rudiments of hinder limbs, which obtain their sustenance from sea-weeds or seaside herbage. They have teeth adapted for bruising such substances, and the movements of the head in grazing require the cervical vertebræ to be unanchylosed; these are, however, short, and in the manatee but six in number. In the dugong (Fig. 26), one of these herbivorous sea-mammals frequenting the Malayan and Australian shores, the upper and lower jaws are singularly bent down, and the upper jaw is armed with a pair of short tusks. The bones of all these cetacea are singularly massive and compact. Three or four of the anterior thoracic ribs are joined to a sternum—the rest are free. One of the vertebræ intervening between the costal and caudal series has connected with it a simple pelvic arch, in which the ilium and ischium may be recognized, and a still more rudimental condition of such arch is suspended in the inguinal muscles of the true cetacea. Most of the caudal vertebræ (Fig. 26), *cd*, of the

Fig. 26.

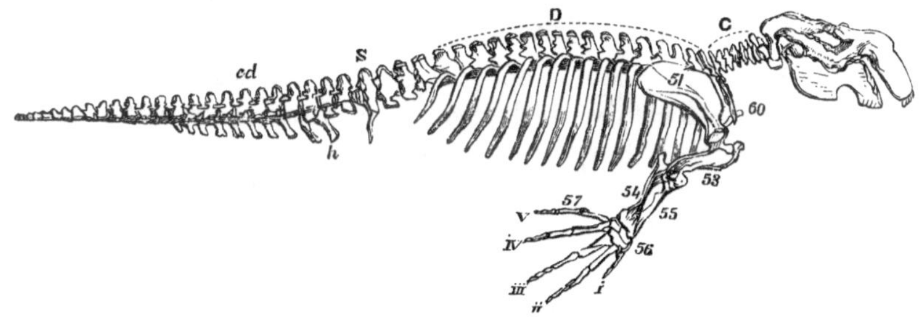

SKELETON OF THE DUGONG (*Halicore Australis*).

manatee and dugong, have long diapophyses, and hæmal arches (Fig. 26), *h*. The terminal vertebræ are flattened horizontally.

The lacteal organs of the dugong are placed on the breast, and the pectoral fins, in the female at least, are

14

applied to clasp the young; and the animal so observed, with its own head and that of its young above water, has given rise to the fable of the siren and mermaid. The bones and joints of the pectoral fin are accordingly better developed than in the ordinary whales. The first row of carpal bones, 56, consists of two—one articulated to the radius, 55, the other to the ulná, 54, and fifth digit, 57, *v*, and both to the single bone representing the second row. The first digit, *i*, consists of a short metacarpal; the metacarpals of the others support each three phalanges.

SKELETON OF THE SEAL.

In the seal tribe (*Phocidæ*), another and well-marked stage is gained in the development of the terrestrial instruments of locomotion. Hind limbs are now added—the marine mammal has become a quadruped. The sphere of life of the seals is near the shores; they often come on land; they sleep and bring forth among the rocks and littoral caves: hence the necessity for a better development of the pectoral limbs, although these, like the pelvic ones, still retain the general form of fins. The fish-hunting seals make more use of the head in independent movements of sudden extension, retraction, and quick turns to the right and left, than do the cetacea of like diet; and the walrus (Fig. 27) works the head, as the place of attachment of its long, vertical, down-growing tusks, in various movements required in climbing over floes and bergs of ice. Accordingly, in the seal tribe, we find the seven neck-vertebræ (*ib.*) *c*, longer, and with more finished and free-playing joints than in the whales and dugongs. The sigmoid curve, in which they can be thrown during retraction of the head, exceeds that in most other mammals, and almost reminds one of the extent of flexion of this part of the spine in birds.

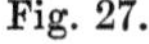

Fig. 27.

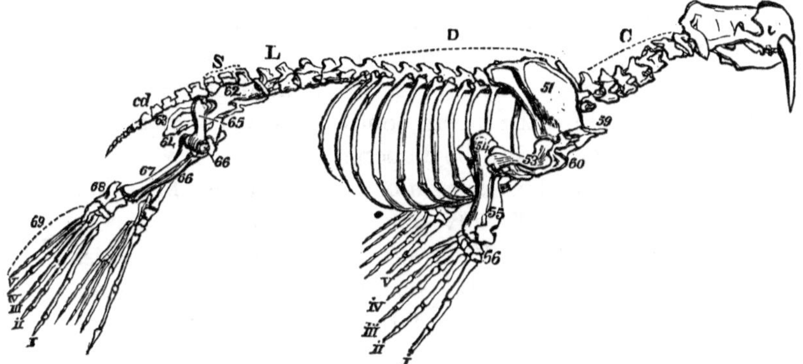

SKELETON OF THE WALRUS (*Trichecus rosmarus*).

In the walrus, the skeleton of which is here selected to exemplify the phocal modification of the mammalian skeleton, the vertebral formula is: 7 cervical, C, 11 dorsal, D, 5 lumbar, L, 3 sacral, S, and 9 caudal, *cd.* As, in consequence of the presence of hind-limbs, a sacrum is now established, the characters of the above five kinds of body-vertebræ, as defined in man and other mammals, may here be given: the cervical or neck-vertebræ "have perforated transverse processes," the dorsal vertebræ "bear ribs;" the lumbar vertebræ "have imperforate transverse processes and no ribs;" the sacral vertebræ "are anchylosed together;" the rest are caudal vertebræ, whatever their modifications. In the above characters, the term "rib" is given to the vertebral element called "pleurapophysis," when this is long and movable; that element may be, and often is, present, but short and fixed, in both cervical, lumbar, sacral, and caudal vertebræ; in some mammals, *e. g.* monotremes, the pleurapophysis may remain unanchylosed in some of the neck-vertebræ, but it is short, like a transverse process; and the so-called "perforated transverse process" in all mammals consists of the diapophysis, parapophysis, and pleurapophysis; the hole being the interval between those parts; in the lumbar vertebræ the pleurapophysis is short, and confluent or connate with the diapophysis.

Returning to the skeleton of the walrus, we find that nine pairs of ribs directly join the sternum, which consists of eight bones. The transverse processes of the last cervical are imperforate, consisting of the diapophysis only. The neural arches of the middle dorsal vertebræ are without spines and very narrow, leaving wide unprotected intervals of the neural canal. The bones of the neck are modified to allow of great extent and freedom of inflection. The perforated transverse processes of the third to the sixth cervicals inclusive are remarkable for the distinctness of their constituent parts. Inferior ridges and tuberous processes, called "hypapophyses," are developed from some dorsal and lumbar vertebræ. These processes indicate the great development of the anterior vertebral muscles, *e. g.* the "longi colli" and "psoæ," and relate to the important share which the vertebræ and muscles of the trunk take in the locomotion of the sea-tribe, especially when on dry land, where they may be called "gastropods," in respect of their peculiar mode of progression. The walrus alone seems to have the power of supporting itself on the fore fins, so as to raise the belly from the ground. There is no trace of clavicle in any seal. The upper part of the scapula exceeds the lower one in breadth. The spine terminates by a short and simple acromion. The humerus is short and thick, and is remarkable for the great development of the inner tuberosity and of the deltoid ridge, which is deeply excavated on its outer side. The inner condyle is perforated. The scaphoid and lunar bones are connate. Although the pollex or the first digit exceeds the third, fourth, and fifth in length, it presents its characteristic inferior number of phalanges, by which the front border of the fin is rendered more resisting. The pelvic arch is remarkable for the stunted development of the ilia, and the great length of the ischia and pubes. The femur is equally peculiar for its shortness and breadth. The tibia and fibula present the more usual proportions, and are anchylosed at their proximal ends. The bones of the foot are strong, long, and are modified to form the basis of a large and powerful fin: the middle toe is the shortest, and the rest increase in length to the margins of the foot; the inner toe has, nevertheless, but two phalanges, the rest having three phalanges, whatever their length; and this is the typical character, both as to the number of the digits and their joints, in both fore and hind feet of the mammalia.

In the living walrus and seal, the digits of each extremity are not only bound together by a common broad web of skin, but those of the hind-limbs are closely connected with the short tail: being stretched out backwards, they seem to form with it one great horizontal caudal fin, and they constitute the chief locomotive organ when the animal is swimming rapidly in the open sea. The long bones of seals, like those of whales, are solid.

With regard to the skull in the seal-tribe, it may be remarked that an occipitosphenoidal bone is formed, as in man, by the coalescence of the basioccipital with the basisphenoid; the parts of the dura mater or outer membrane of the brain, called "tentorium," with the posterior part of the "falx," are ossified. The sella turcica is shallow, but well defined behind by the overhanging posterior clinoid processes: the petrosal shows a deep, transverse, cerebellar fossa, and is perforated by the carotid canal. The frontal forms a small rhinencephalic fossa, and contributes a very large proportion to the formation of the orbital and olfactory chambers.

In Fig. 27, 62 is the ilium, 63 the ischium, and 64 the pubes, 65 is the femur or thigh-bone, 66 the tibia, 66′ the patella or kneepan, 67 the fibula, 68 the tarsus, and 69 the metatarsus and phalanges of the hind-foot; the numbers on the other bones correspond with those in the skeleton of the dugong.

SKELETONS OF HOOFED QUADRUPEDS—THE HORSE.

The contrast, as regards the sphere of life and kind of movement between the seal and the horse is very great; the instruments of locomotion, and indeed the whole frame, need to be very different in an animal that can only shuffle on its belly along the ground, and one that can traverse the surface of the earth at the rate of four miles in six minutes and a half, as was achieved by the noted racer "Flying Childers." The modifications in the form and proportions of the locomotive members are accordingly extreme. The limbs in the horse are as remarkable for their length and slenderness, as in the seal for their brevity and breadth. Both fore and hind limbs in the horse terminate each in a single hoof; the trunk is raised high above the ground, and is more remarkable for its depth than breadth, especially at the fore part; the neck is long and arched; the jaws long and slender, being produced so as to facilitate the act of cropping the grass, and leaving so much space between

Fig. 28.

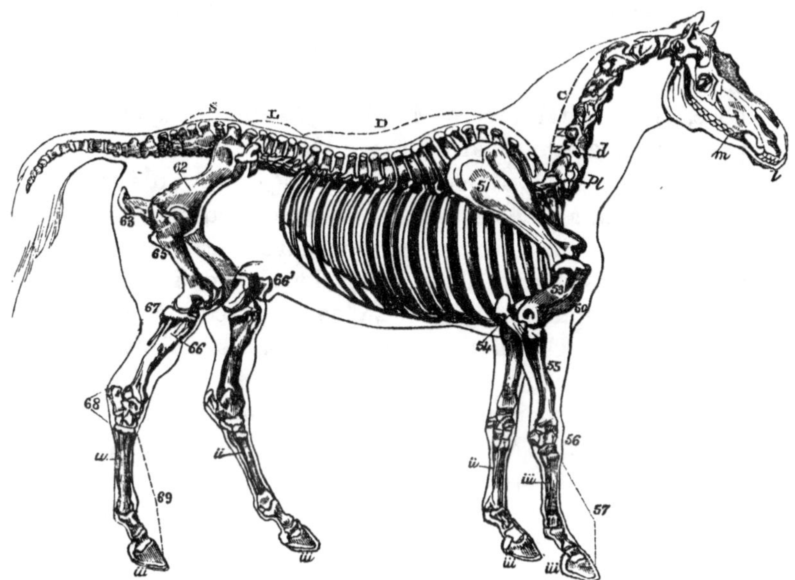

HORSE (*Equus caballus*).

the front teeth, *i*, and the grinders, *m*, as permits man to insert the instrument called "bit" into the mouth, whereby he masters and guides his noble and valuable four-footed ally, as the ship is steered by the helm.

Were every animal constructed expressly and exclusively for its own peculiar habits of life, and irrespective of any common pattern, it could scarcely be expected, beforehand, that the same bones would be found in the horse as in the seal; yet a comparison of their skeletons, Cuts 27 and 28, will demonstrate that this is, to a very great degree, the case.

The vertebral formula of the horse is: 7 cervical, C, 19 dorsal, D, 5 lumbar, L, 5 sacral, S, and 17 caudal. Eight pairs of ribs directly join the sternum, 60, which consists of seven bones and an ensiform cartilage. The neural arches of the last five cervical vertebræ expand above into flattened, subquadrate, horizontal plates of bone, with a rough tubercle in place of a spine: the zygapophyses, *z*, are unusually large. The perforated transverse process sends a pleurapophysis, *pl*, downwards and forwards, and a diapophysis, *d*, backwards and outwards, in the third to the sixth cervicals inclusive: in the seventh the diapophysial part alone is developed, and is imperforate. The spinous processes suddenly and considerably increase in length in the first three dorsals, and attain their greatest length in the fifth and sixth, after which they gradually shorten to the thirteenth, and continue of the same length to the last lumbar. The lumbar diapophyses are long, broad, and in close juxtaposition; the last presents an articular concavity adapted to a corresponding convexity on the fore part of the diapophysis of the first sacral. The scapula, 51, is long and narrow, and according to its length and obliquity of position the muscles attached to it, which act upon the humerus, operate with more vigor, and to this bone the attention of the buyer should be directed, as indicative of one of the good points in a horse. The coracoid is reduced to a mere confluent knob. The spine of the scapula, 51, has no acromion. The humerus, 53, is remarkable for the size and strength of the proximal tuberosities in which the scapular muscles are implanted. The joint between it and the scapula is not fettered by any bony bar connecting the bladebone with the breastbone; in other words, there is no clavicle. The ulna, represented by its olecranal extremity, 54, is confluent with the radius, 55. The os magnum in the second series of carpal bones, 56, is remarkable for its great breadth, corresponding to the enormous development of the metacarpal bone of the middle toe, which forms the chief part of the foot. Splint-shaped rudiments of the metacarpals, answering to the second, *ii*, and fourth, *iv*, of the pentadactyle foot, are articulated respectively to the trapezoides and the reduced homologue of the unciforme. The mid-digit, *iii*, consists of the metacarpal called "cannon-bone," and of the three phalanges, which have likewise received special names in Veterinary Anatomy, for the same reason as other bones have received them in Human Anatomy. "Phalanges" is the "general" term of these bones, as being indicative of the class to which they belong, and "hæmapophyses" is the "general" term of parts of the inferior arches of the head-segments; and just as, from the modifications of these hæmapophyses, they have come to be called "maxilla," "mandibula," "ceratohyal," &c., so the phalanges of the horse's foot are called—the first, "great pastern bone," the second, "small pastern bone," and the third, which supports the hoof, the "coffin bone;" a sesamoid ossicle between this and the second is called the "coronary." The ilium, 52, is long, oblique, and narrow, like its homotype, the scapula; the ischium, 63, is unusually produced backwards. The extreme points of these two bones show the extent to which the bending muscles and extending muscles of the leg are attached; and according to the distance of these points from the thigh-bone the angle at which they are therein inserted becomes more favorable for their force; the longer, therefore, and the more horizontal the pelvis, the better the hind-quarter of the horse, and its qualities for swiftness and maintenance of speed depend much on the "good point" due to the development of this part of the skeleton. The femur, 65, is characterized by a third trochanter springing from the outer part of the shaft before the great trochanter. There is a splint-shaped rudiment of the proximal end of the fibula, 67, but not any rudiment of the distal end. The tibia, 66, is the chief bone of the leg. The heel-bone, "calcaneum," is much produced, and forms what is called the "hock." The astragalus is characterized by the depth and obliquity of the superior trochlea, and by the extensive and undivided anterior surface, which is almost entirely appropriated by the naviculare. The external cuneiforme is the largest of the second series of tarsals, being in proportion to the metatarsal of the large middle digit, *iii*, which it mainly supports. The diminished cuboides articulates partly with this, partly with the rudiment of the metatarsal corresponding with that of the fourth toe, *iv*. A similar rudiment of the metatarsal of the toe, corresponding with that of the second, *ii*, articulates with a cuneiforme medium—here, however, the innermost of the second series of tarsal bones.

Of all the other known existing hoofed quadrupeds, it would hardly be anticipated that the rhinoceros presented the nearest affinity to the horse; one might rather look to the light camel or dromedary; but a different modification of the entire skeleton may be traced in the animals with toes in even number, as compared with the horse and other odd-toed hoofed quadrupeds. In an extinct kind of horse (*Hippopotherium*), the two splint-bones are more developed, and each supports three phalanges, the last being provided with a diminutive hoof. In the extinct *Palæotheria*, the outer and inner digits acquired stronger proportions, and

Fig. 29.

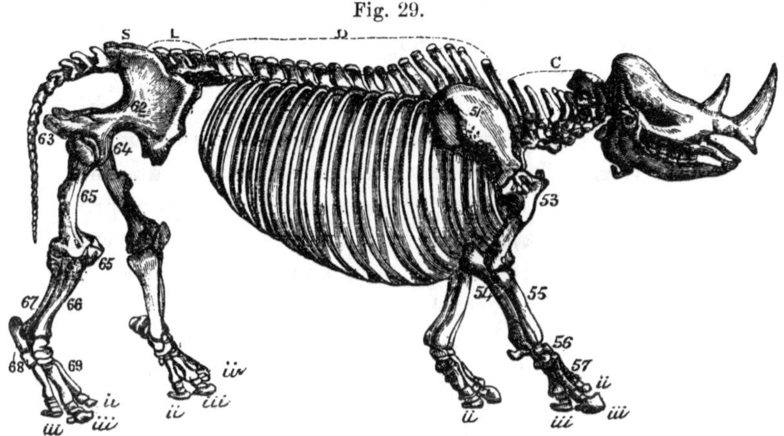

SKELETON OF THE RHINOCEROS (*Rh. bicornis*).

the entire foot was shortened. The transition from the *Palæotheria*, by the extinct hornless rhinoceros (*Acerotherium*), to the existing forms of rhinoceros, is completed. In the skeleton of the rhinoceros, we find resemblances to the horse in the number of the dorsal vertebræ, in the third trochanter of the femur, and in the number of digits on each foot, albeit the two that are hidden and rudimental in the swifter quadruped are here made manifest in their full development: the concomitant shortening of the whole foot, and strengthening of the entire limbs, accord with the greater weight of the body to be supported, clad as it is with a coat-armor of thickened tuberculated hide: the broader feet, terminated each by three hoofs, afford a better basis of support in the swampy localities affected by the rhinoceros. Both scapulæ and iliac bones are of greater breadth, and less length. The ulna is fully developed in the fore-limb, and the fibula in the hind-leg; but there is no power of rotation of the fore-limb in any hoofed quadruped. The upper surface of the skull is roughened for the attachment of the horn, and in two distinct places where the species has two horns.

If the equine skull be compared with that of the rhinoceros, the basioccipital will be seen to be narrow and more convex. The true mastoid intervenes, as a tuberous process, between the post-tympanic and paroccipital process, clearly indicating the true nature of the post-tympanic in the rhinoceros; the tapir shows an intermediate condition of the mastoid between the rhinoceros and horse. The latter differs from both the tapir and rhinoceros in the outward production of the sharp roof of the orbit and the completion of the bony frame of that cavity behind by the junction of the postorbital process with the zygoma. The temporal fossa, so defined, is small in proportion to the length of the skull: the base of the postorbital process is perforated by a superorbital foramen. The lachrymal canal begins by a single foramen. The premaxillaries extend to the nasals, and shut out the maxillaries from the anterior aperture of the nostrils. The chief marks of affinity to other odd-toed hoofed beasts (*Perissodactyles*) are seen in the shape, size, and formation of the posterior aperture of the nostrils, the major part of which is bounded by the palatine bones, of which only a small portion enters into the formation of the bony palace, which terminates behind opposite the interspace between the penultimate and last molars. A narrow groove divides the palato-pterygoid process from the socket of the last molar, as in the tapir and rhinoceros. The pterygoid process has but little antero-posterior extent: its base is perforated by the ectocarotid canal. The entopterygoids are thin plates, applied like splints over the inner side of the squamous suture between the pterygoid processes of the palatines and alisphenoids. The postglenoid process in the horse is less developed than in the tapir. The Eustachian process is long and styliform. There is an anterior condyloid foramen, and a wide "fissura lacera." The broad and convex bases of the nasals articulate with the frontals a little behind the anterior boundary of the orbits. The space between the incisors and molars is of greater extent than in the tapir; a long diastema is not, however, peculiar to the horse; and, although it allows the application of the bit, that application depends rather upon the general nature of the horse, and its consequent susceptibility to be broken in, than upon a particular structure which it possesses in common with the ruminants and some other herbivora.

The tapir and the rock cony have four digits on each fore-foot, and three digits on each hind-foot; but they resemble more the horse and rhinoceros than any other *Ungulata.* If the osteological characters of the hoofed animals with the hind digits in uneven number be compared together, they will be found to present, notwithstanding the differences of form, proportion, and size presented by the rhinoceros, hyrax, tapir, and horse, the following points of agreement, which are the more significative of natural affinity when contrasted with the skeletons of the hoofed animals with digits in even number. Thus, in the odd-toed or "perissodactyle" ungulates, the dorso-lumbar vertebræ differ in different species, but are never fewer than twenty-two; the femur has a third trochanter, and the medullary artery does not penetrate the fore part of its shaft. The fore part of the astragalus is divided into two very unequal facets. The os magnum and the digitus medius which it supports is large, in some disproportionately, and the digit is symmetrical; the

Fig. 30.

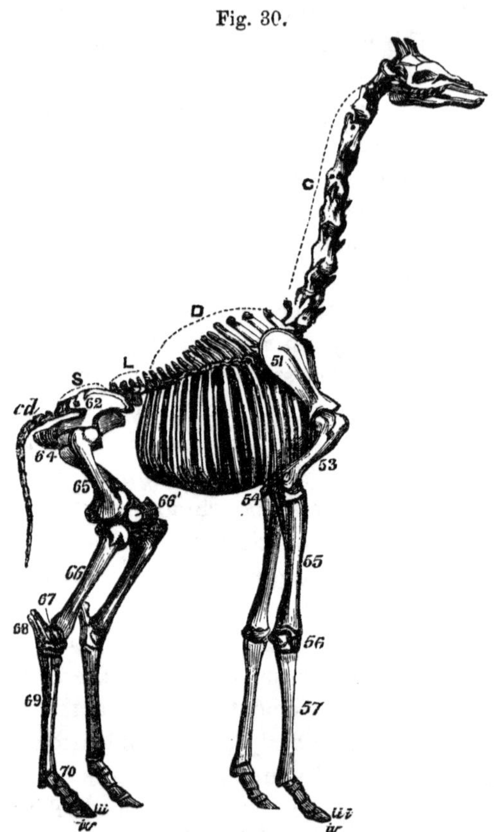

SKELETON OF THE GIRAFFE (*Camelopardalis giraffa*).

same applies to the ectocuneiform and the digit it supports in the hind foot. If the species be horned, the horn is single; or if there be two, they are placed on the median line of the head, one behind the other, each being thus a single or odd horn. There is a well-developed post-tympanic process, which is separated by the true mastoid from the paroccipital in the horse, but unites with the lower part of the paroccipital in the tapir, and seems to take the place of the mastoid in the rhinoceros and hyrax. The hinder half, or a larger proportion, of the palatines enters into the formation of the posterior nares, the oblique aperture of which commences in advance of the last molar, and, in most, of the penultimate one. The pterygoid process has a broad and thick base, and is perforated lengthwise by the ectocarotid. The crowns of the antepenultimate, as well as the penultimate and last premolars, are as complex as those of the molars; that of the last lower milk-molar is bilobed. To these osteological and dental characters may be added some important modifications of internal structure, as, *e. g.*, the simple form of the stomach, and the capacious and sacculated cæcum, equally indicating the mutual affinities of the odd-toed or perissodactyle hoofed quadrupeds, and their claims to be regarded as a natural group of the Ungulata. Many extinct genera, *e. g.*, lophiodon, tapiotherium, palæotherium, hippotherium, acerotherium, macrauchenia, elasmotherium, coryphodon, have been discovered, which once linked together the now broken series of Perissodactyla, represented by the existing genera rhinoceros, hyrax, tapyrus, and equus.

Another series of hoofed quadrupeds is characterized by having their hoofs and digits in even number in both fore and hind feet. The majority of these have a pair, so developed as to serve as feet, and terminated by a pair of hoofs so shaped as to look like one split hoof, whence the name "cloven-footed," given to this predominant family of "artiodactyle," or even-toed beasts; the synonym "ruminant," indicative of the same great family, is deduced from the characteristic complexity of their act of digestion.

No food is more remote or distinct from flesh than grass. Extremities enveloped in hoofs are incapacitated from seizing and retaining a living prey, hence all hoofed mammals are necessarily herbivorous: hence the complexity of their grinding teeth, the concomitant strength of their grinding muscles, and weakness of the biting muscles; the length of the neck, to enable the head to reach the verdant earth, and the length and slenderness of the jaws. The absence of a clavicle, and of any power of rotating the bones of fore-leg and fore-foot, are also constant characteristics of both great divisions of the *Ungulata* or hoofed quadrupeds.

The ox, the hog, and the hippopotamus are examples of even-toed hoofed quadrupeds. In the ox, besides the two large and normally developed hoofs, two small supplementary hoofs dangle behind, in each foot; in the hog these are brought down to the level of the midpair, but are smaller; in the hippopotamus the four digits and hoofs are subequal on each foot. From this type of extremity to that of the giraffe, or camel, where the digits are absolutely restricted to two on each foot, there is a close series of gradational shortcomings affecting the outer and the inner toes, until they wholly disappear. The giraffe (Fig. 30) is a ruminant dwelling in climes where herbage disappears from the parched soil soon after the rainy season has terminated, and where sustenance for a herbivore of its bulk could hardly be afforded, except by trees: it is therefore modified to browse on the tender branches, and chiefly of the light and lofty acacias. Its trunk is accordingly short, and raised high upon long and slender limbs, especially at the forepart; a small and delicate head is supported on an unusually long neck. The number of vertebræ here, however, accords with that characteristic of the mammalian class, viz: seven. They are peculiar for the length of their bodies. There are fourteen dorsal vertebræ with very long spinous processes, and supporting long and slender ribs, especially the anterior ones, seven pairs of which join the sternum, which consists of six bones; the lumbar vertebræ are five in number, the sacral four, and the caudal twenty; this series is terminated in the living animal by a tuft of long, wavy, stiff black hair, forming an admirable whisk to drive off insect tormentors. The bladebone, 51, is remarkably long and slender; its spine or ridge forms a very low angle, and gradually subsides as it approaches the neck of the scapula; the coalesced coracoid is a large tuberosity. The humerus, 53, forms the shortest of the three segments of the limb; it is remarkable for the strength of the proximal processes; the second segment is chiefly constituted, as in all ruminants, by the radius, 55; the slender shaft of the ulna, 56, which supports a long olecranon, becomes blended with the radius, and lost at its lower third, but its distal end reappears as a distinct part. The metacarpals of the retained digits, answering to the third and fourth in the human and other five-fingered (pentadactyle) hands, are blended together to form a single "cannon-bone" of the veterinarians; but the nature of this is different from that in the horse; it divides at its distal end into two well-formed trochleæ, or pulley-joints, and to these are articulated the digits *iii* and *iv*, which each consist of three joints or phalanges. Thus the main extent of this singularly elongated limb is gained by the excessive development of the hand-segment, restricted, however, to those elements that answer to the middle and ring-fingers of the human hand.

The pelvis, of which the sacral, S, iliac, 62, and ischial, 64, elements are shown in Cut 31, is small in proportion to the animal's bulk. The femur, 65, is short like the humerus, and chiefly remarkable for the great expanse of this distal end. The tibia, 66, forms the main basis of the leg, as its homotype the radius does in the forearm, but the fibula is more reduced than in the ulna; rarely in any ruminant is more of it visible than its distal end, 67, wedged in between the tibia and the calcaneum. The series of tarsal bones, 68, is peculiar in all ruminants for a coalescence of the two bones answering to the "scaphoid and cuboid" in the human tarsus. In all ruminants the astragalus is unusually symmetrical in shape, with a deep trochlear articular surface for the tibia, and two equal convex surfaces for succeeding tarsal bones; the calcaneum is produced into a long "hock." The rest of the bones of the hind-foot conform closely with those of the fore-foot.

A few remarks, although interesting chiefly to the professed anatomist, appear called for in reference to the bony structure of the head of the giraffe.

The exoccipitals form a marked protuberance above the foramen magnum, and below a deep fossa, for the implantation of the ligamentum nuchæ—the length of the dorsal spines being related, in all ruminants, to a due surface for the attachment of this strong elastic support of the head and neck. The parietals are chiefly situated on the upper surface of the skull; the osseous horn-cores, which are originally distinct, become anchylosed in old giraffes, across the coronal suture, equally to the parietals and frontals; if one of these be divided longitudinally, it will show the extension of the frontal and parietal sinuses into its lower fourth, the rest of the horncore being a solid and dense bone. The protuberance upon the frontal and contiguous parts of the nasal bones is entirely due to an enlargement of those bones, and not to any distinct osseous part: its surface is roughened by vascular impressions. The lachrymal is separated from the nasal by a large vacuity intervening between those bones, the frontal and the maxillary. The premaxillaries, which are of unusual length, articulate with the nasals. The petrotympanic is a separate bone, as in all ruminants. The symphysis of the lower jaw is unusually long and slender in the giraffe.

In the skeleton of the Camel (*Camelus bactrianus*) the vertebral formula is—seven cervical, twelve dorsal, seven lumbar, four sacral, eighteen caudal. Seven pairs of ribs articulate directly with the sternum, which consists of six bones, the last being greatly expanded and protuberant below, where it supports the pectoral callosity in the living animal. The cervical region, though less remarkable for its length than in the giraffe, is longer than in ordinary ruminants, and is remarkable for its flexuosity; the vertebræ are peculiar for the absence of the perforation for the vertebral artery in the transverse process, with the exception of the atlas; that artery, in the succeeding cervicals, enters the back part of the neural canal, and perforates obliquely the forepart of the base of the neurapophysis. The costal part of the transverse process is large and lamelliform in the fourth to the sixth cervical vertebræ inclusive: in the seventh it is a short protuberance. The spinous process of the first dorsal suddenly exceeds in length that of the last cervical, and increases in length to the third dorsal; from this to the twelfth dorsal the summits of the spines are on almost the same horizontal line, and are expanded and obtuse above, sustaining the substance of the two humps of this species; they afford, however, no other indication of those risings, which are as independent of the osseous system as is the dorsal fin in the grampus or porpoise. The spines of the lumbar vertebræ progressively decrease in length. The spine of the scapula is produced into a short-pointed acromion: the coracoid tubercle is large, and grooved below. The ridge upon the outer condyle of the humerus is much less marked than in the normal ruminants. The ulna has coalesced more completely with the radius, and appears to be represented only by its proximal and distal extremities. The carpal bones have the same number and arrangement as in ordinary ruminants, but the pisiforme is proportionally larger. There is no trace of the digits answering to the first, second, and fifth in the pentadactyle foot: the metacarpals of those answering to the third and fourth have coalesced to near their distal extremities, which diverge more than in the ordinary ruminants, giving a greater spread to the foot, which is supported by the ordinary three phalanges of each of those digits. The last phalanx deviates most from the form of that in the ordinary ruminants by its smaller proportional size, rougher surface, and less regular form: it supports, in fact, a modified claw rather than a hoof. In the femur the chief deviation from the ordinary ruminant type is seen in the position of the orifice of the canal for the medullary artery, which, as in the human skeleton, enters the back part of the middle of the shaft, and inclines obliquely upwards. The fibula is represented by the irregularly-shaped ossicle interlocked between the outer side of the distal end of the tibia and the calcaneum. The scaphoid is not confluent with the cuboid as in the normal ruminant: the rest of the hind-foot deviates in the same manner and degree from the ordinary ruminant type, as does the fore-foot.

The camel tribe have no horns; some small deer of the musk-family are compensated for the want of horns by very long and sharp upper canine teeth; the rest of the ruminants, either in the male sex or in both sexes, are endowed with the weapons of offence and defence, developed from and supported by the head, called "horns" and "antlers." The term "horn" is technically restricted to the weapon which is composed of a bony base, covered by a sheath of true horny matter. Such horns are never shed; and as, in order to diminish the weight of the head, the horn-core is made as hollow as is consistent with strength, the ruminants with such horns are called hollow-horned; the ox, the sheep, and the antelope are examples. Antlers consist of bone only. During the period of their growth they are covered by a vascular, short-haired skin like velvet; but when their growth is completed, this skin dries and peels off, leaving the antler a solid, naked, and insensible weapon. Being deprived, however, of its vascular support it dies, and, after a certain period of service, is undermined by the absorbents and cast off. The process of growth and decadence of the antlers is repeated each year; and in the fallow-deer the antlers progressively acquire greater size and more branches to the sixth year, when the animal is in its prime. Good evidence has been obtained that the same law of growth, shedding, and annual renewal prevailed in the gigantic fossil deer of Ireland, in which upwards of eighty pounds of osseous matter must have been developed from the frontal bones every year in the full-grown animal. The ruminants of the deer and elk tribes are those which have antlers, or are "solid-horned." The horns of the giraffe are peculiar; they are short and simple, are always covered by a hairy integument, and are never shed. They relate in position to both the frontal and parietal bones. In all other ruminants, the horns are developed from the frontals exclusively, although they sometimes, as in the ox, project from the back part of the cranium; but the frontals, in such cases, extend to that part. The horn of the rhinoceros consists wholly of fibrous horny matter.

The even-toed hoofed animals that do not ruminate have no horns. The osteology of this division is here illustrated in the hippopotamus (Fig. 31). The skeleton,

Fig. 31.

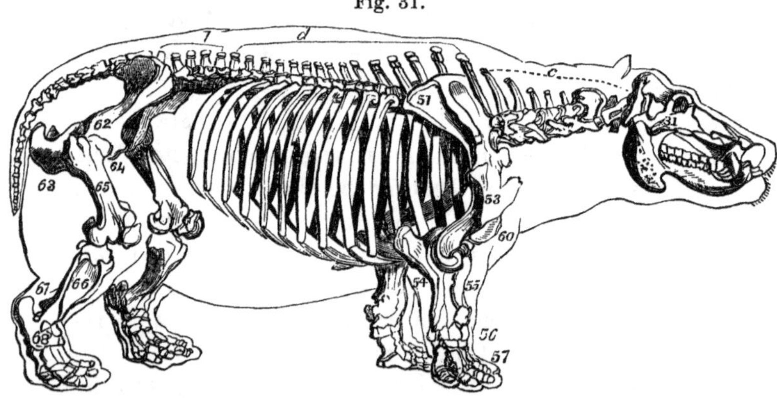

SKELETON OF THE HIPPOPOTAMUS AMPHIBIUS.

in its strength and massiveness, presents a greater contrast with that of the giraffe than the rhinoceros's skeleton does with that of the horse; there are, nevertheless, as will be shown in the concluding summary, more essential points of resemblance to the giraffe's skeleton than to that of the rhinoceros. In points of minor importance, we find the hippopotamus resembling the rhinoceros; as *e. g.* in the shortness and strength of its neck; but it has only fifteen dorsal, *d*, and four lumbar, *l*, vertebræ. The spines of these vertebræ are shorter and less unequal than in the ruminants; and they have an almost uniform direction, as in all quadrupeds that do not move by leaps or bounds. The tail is short, and, in the living animal, compressed, acting like a rudder. The bones of the limbs are short and thick. In the scapula, 51, the acromion is slightly produced, and the coracoid recurved. The great tuberosity of the humerus, 53, is divided into two subequal processes. The ulna and radius have coalesced at their extremities, and at the middle of their shaft, the interosseous space being indicated by a deep groove and two holes. In the carpal series of bones, the trapezium is present, but does not support any digit; the innermost, answering to the thumb or pollex, therefore, is the one which is absent; of the remaining four digits, the two middle ones, answering to the third and fourth, are most developed. The femur has no third trochanter. The fibula is distinct from the radius, and extends from its proximal end to the calcaneum. The entocuneiform bone is present in the tarsus, but there is no rudiment of the innermost toe or hallux; the proportions of the other four toes resemble those on the fore-foot.

The skull is remarkable for the prominence and high position of the orbits, which allow the eye to be projected above the surface of the water, and a survey to be made by the suspicious animal without the exposure of any other part of the head. The upper jaw is peculiar for the development of the sockets of the great canine teeth, and the lower jaw combines with the like character an unusual production and curvature of the angle.

With regard to the osteology of the hog tribe, our limits compel us to restrict ourselves to the notice of the still more singular development of the sockets of the upper canines or tusks in the babyroussa, in which those teeth curve upwards and pierce the skin of the face, like horns, whence the name "horned hog" sometimes imposed upon it.

If the hoofed animals with the digits in even number be compared together, in regard to their osteological characters, they will be found, notwithstanding the difference of form, proportion, and size presented by the hippopotamus, wild boar, vicugna, and chevrotain, to agree in the following points, which are the more significative of natural affinity when contrasted with the skeletons of the hoofed animals with digits in uneven number. Thus, in the even-toed or "artiodactyle" ungulates, the dorso-lumbar vertebræ are the same in number, as a general rule, in all the species, being nineteen. The rare exceptions appear to be due to the development, rarely to the suppression, of an accessory vertebra as an individual variety, the number in such cases not exceeding twenty, or falling below eighteen, and the supernumerary vertebra being most usually manifested in the domesticated and highly-fed breeds of the common hog. The recognition of this important character appears to have been impeded by the variable number of movable ribs in different species of the artiodactyles, the dorsal vertebræ, which these ribs characterize, being fifteen in the hippopotamus and twelve in the camel; and the value of this distinction has been exaggerated owing to the common conception of the ribs as special bones, distinct from the vertebræ, and their non-recognition as parts of vertebra equivalent to the neurapophyses and other autogenous elements. The discovery of the pleurapophyses under the condition of rudimental ribs attached to the ends of the lumbar diapophyses, which afterwards become suturally attached or anchylosed, and the pleurapophysial nature of a part of the so-called perforated transverse process of the cervical vertebra, show that the anthropotomical definition of a dorsal vertebra, as one that supports ribs, is inapplicable to the mammalia generally, and is essentially incorrect. It is convenient, in comparative tables of vertebral formulæ, to denote the number of those vertebræ of the trunk in which the pleurapophyses remain free and movable, constituting the "ribs" of anthropotomy; but the differences sometimes occurring in this respect, in individuals of the same species, have their unimportance manifested when the true nature of a rib is recognized. The vertebral formulæ of the artiodactyle skeletons show that the difference in the number of the so-called dorsal and lumbar vertebræ does not affect the number of the entire dorso-lumbar series: thus, the Indian wild boar has *d*, 13, *l*, 6 = 19; *i. e.* 13 dorsal, and 6 lumbar, making a total of 19 trunk-vertebræ; the domestic hog and the peccari have *d*, 14, *l*, 5, = 19; the hippopotamus has *d*, 15, *l*, 4, = 19; the gnu and aurochs have *d*, 14, *l*, 5, = 19; the ox, and most of the true ruminants, have *d*, 13, *l*, 6, = 19; the camel and lamas have *d*, 12, *l*, 7, = 19. These facts illustrate the natural character and true affinities of the artiodactyle group. They are further shown by the absence of the third trochanter in the femur, and by the place of perforation of the medullary artery at the fore and upper part of the shaft, as in the hippopotamus, the hog, and most of the ruminants. The fore part of the astragalus is divided into two equal or subequal facets; the os magnum does not exceed, or is less than, the unciforme in size, in the carpus; and the ectocuneiforme is less, or not larger, than the cuboid, in the tarsus. The digit answering to the third in the pentadactyle foot is unsymmetrical, and forms, with that answering to the fourth, a symmetrical pair. If the species be horned, the horns form one pair or two pairs; they are never developed singly and symmetrically from the median line. The post-tympanic does not project downward distinctly from the mastoid, nor supersede it in any artiodactyle; and the paroccipital always exceeds both in length. The bony palate extends further back than in the perissodactyles; the hinder aperture of the nasal passages is more vertical, and commences posterior to the last molar tooth. The base of the pterygoid process is not perforated by the ectocarotid artery. The crowns of the premolars are smaller and less complex than those of the true molars, usually representing half of such crown. The last milk-molar is trilobed.

To these osteological and dental characters may be added some important modifications of internal structure, as *e. g.* the complex form of the stomach in the hippopotamus, peccari, and ruminants, the comparatively small and simple cæcum, and the spirally folded colon, which equally indicate the mutual affinities of the even-toed or artiodactyle hoofed quadrupeds, and their claims to be regarded as a natural group of the Ungulata. Many extinct genera, *e. g.* chæropotamus, anthracotherium, hyopotamus, dichodon, merycopotamus, xiphodon, dichobune, anoplotherium, have been discovered, which once linked together the now broken series of Artiodactyla, represented by the existing genera hippopotamus, sus, dicotyles, camelus, moschus, camelopardalis, cervus, antelope, ovis, and bos.

As we have now traced both the fore and hind foot to the five-toed or pentadactyle structure, with the definite number of joints or phalanges in each toe, characteristic of the highest class of vertebrate animals, a few remarks will be offered in illustration of the plan of structure which prevails in such extremities, and of the law that governs the departure from the pentadactyle type in the mammalia.

The essential nature of the limbs is best illustrated by the fish called protopterus, and by some of the lower reptiles that retain gills with lungs.

If the segment of the skeleton supporting the rudiments of the fore-limbs in the protopterus (Fig. 32), be

Fig. 32.

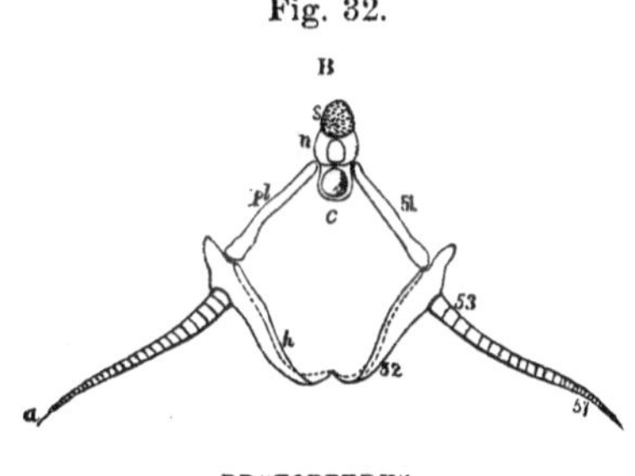

PROTOPTERUS.

compared with the modification of the typical vertebra, exemplified in Fig. 6, p. 28, it will be seen to be constructed on the same type. The hæmal arch is most expanded, and it is composed of a pleurapophysis or vertebral rib, *pl*, and a hæmapophysis or sternal rib, *h*, on each side; the hæmal spine, or sternum, is not here developed; the long, many-jointed ray, *a*, answers to the more simple diverging appendage, *a*, in Fig. 5.

The segment supporting these appendages, or first rudiments of the fore-limb in the fish, is the occipital one, or the last vertebra of the skull. The pleurapophysis of this segment is the seat of all those modifications which have earned for it the special name of "scapula," 51; the hæmapophysis is the seat of those that have led to its being called "coracoid," 52.

The corresponding segment of the batrachian amphiuma (Fig. 33) yields the next important modification of these parts. The scapulæ, *pl*, 51, are detached from the occiput, or neutral arch; the coracoids, *h*, 52, are much expanded; three segments of the diverging appendage, *a*, are ossified, and two of these segments are bifid, showing

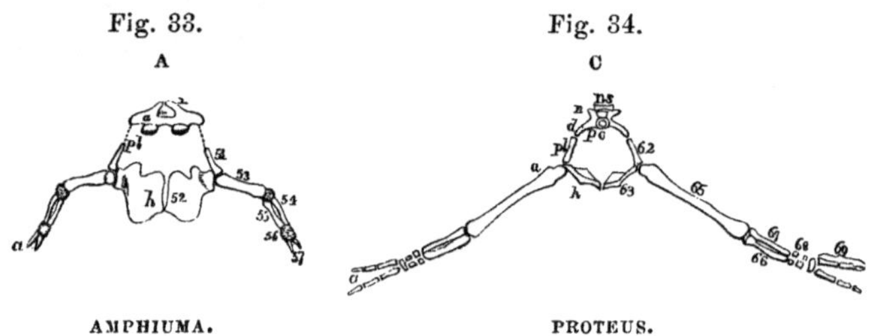
Fig. 33. A — AMPHIUMA. Fig. 34. C — PROTEUS.

the beginning of the radiating multiplication of its parts. The first segment is the seat of those modifications which have obtained for it the special name of "humerus," 53; the two divisions of the next segment of the appendage are called "ulna," 53, and "radius," 54; the remainder of the limb is called "manus," or hand; 56 is the gristly carpus, and the two bony divisions are the digits or fingers, 57.

The segment supporting the hind-limbs retains most of its typical character in the subterranean reptile called the Proteus; one sees, *e. g.* in Fig. 34, that the centrum has coalesced with the neurapophyses, *n*, and neural spine, *ns*, forming the neural arch from which the diapophyses, *d*, are developed: the more expanded hæmal arch consists of the pleurapophyses, *pl*, and the hæmapophyses, *h;* the former is called the "ilium," 62, the latter the "ischium," 63; and, as the hæmapophyses of another segment are usually added to the scapular arch, when they receive the name of "clavicles," so also the hæmapophyses of a contiguous segment are usually added to the pelvic arch, when they are called "pubic bones."

The pelvic diverging appendage, *a a*, has advanced to the same stage of complexity in the proteus, as the scapular one in the amphiuma; the first ossified segment is called "femur," 65; the divisions of the next segment are respectively termed "tibia," 66, and "fibula," 67; the first set of short bones in the "pes," or foot, are called "tarsals," 68; those of the two toes are called "metatarsals" and "phalanges," 69.

The tarsal bones, from the degree of constancy of their forms and relative positions, have received distinct names. In Fig. 35 of the bones of the hind-foot in the horse, *a* marks the "astragalus," *cl*, the "calcaneum," or heel-bone, the prominent part of which forms the "hock;" *s* is the "scaphoides," or naviculare, *b*, the "cuboid," *ce*, "ectocuneiform," and *cm*, the "mesocuneiform." Now, the ectocuneiform in all mammalia supports the third or middle of the five toes when they are all present, the mesocuneiform supports the second toe, and the cuboides the fourth and fifth. We see, therefore, in the horse, that the very large bone articulated to the ectocuneiform, *ce*, is

Fig. 35. Fig 36.

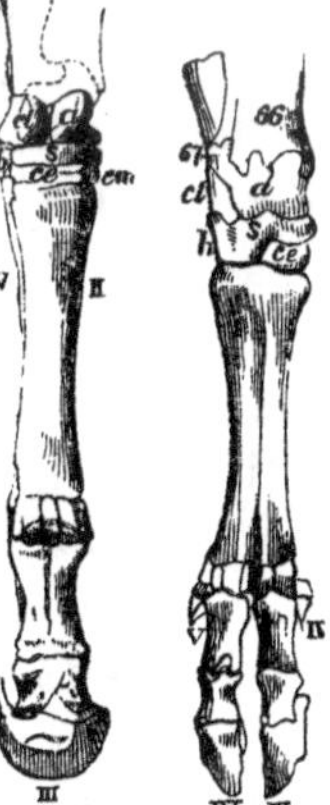

HORSE. OX.

the metatarsal of the *third* toe, to which are articulated the three phalanges of the same toe, *iii*, the last phalanx being expanded to sustain the hoof. The small bone called "split-bone," by veterinarians, articulated to the "meso-cuneiform," is the stunted metatarsal of the *second* toe, *ii;* the outer "splint-bone," articulated to the "cuboides," is the similarly stunted metatarsal of the *fourth* toe, *iv.*

In the foot of the ox (Fig. 36), the cuboides, *b*, presents a marked increase of size, equalling the ectocuneiform, *cc*, which is proportionally diminished. The single bone, called "cannon-bone," which articulates with both these carpal bones, does not answer to the single "cannon-bone" in the horse, but to the metatarsals of both the *third* and the *fourth* digits; it is accordingly found to consist of those two distinct bones in the fœtal ruminant, and there are a few species in which that distinction is retained. Marks of the primitive division are always perceptible, especially at its lower end, where there are two distinct joints or condyles, for the phalanges of the third, *iii*, and fourth, *iv*, toes. In the horse, the rudiments of the two stunted toes were their upper ends or metatarsal bones; in the ox, they consist of their lower ends or phalanges; these form the "spurious hoofs," and are parts of the second, *ii*, and fifth, *v*, toes (Fig. 36). The rhinoceros more closely resembles the horse in the bony structure of its hind-foot (Fig. 37); the ectocuneiform is still the largest of the three lowest tarsal bones, although the mesocuneiform, *cm*, and the cuboids, *b*, have gained increased dimensions in accordance with the completely developed toes which they support; these toes we therefore recognize as being answerable to the rudiments of the second, *ii*, and fourth, *iv*, in the horse, the principal toe being still the third, *iii*. The hippopotamus (Fig. 38) chiefly differs from the ox, as the rhinoceros differs from the horse, viz: by manifesting the two toes fully developed, which were rudimental in the more simple foot; the cuboides, *b*, being proportionally extended to support the fifth toe, *v*, as well as the fourth, *iv;* the second toe, *ii*, articulates, as usual, with a distinct tarsal bone. In the elephant (Fig. 39), where a fifth digit is added,

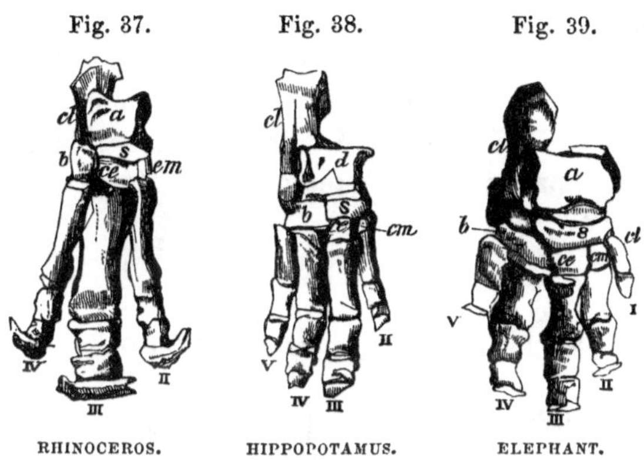

Fig. 37. RHINOCEROS. Fig. 38. HIPPOPOTAMUS. Fig. 39. ELEPHANT.

answering to our first or great toe, I, there is also a distinct carpal bone, called the "entocuneiform," *ci*, and the tarsus presents, as in other pentadactyle mammals, all the bones which are seen in the human tarsus, viz: the astragalus, *a*, the calcaneum, *c*, the scaphoides, *s*, the entocuneiform, *ci*, the mesocuneiform, *cm*, the ectocuneiform, *cc*, and the cuboides, *b*.

The course of the simplification of the pentadactyle foot or hand is first a diminution and removal of the innermost digit, *i;* next of the outermost, *v;* then of the second, *ii;* and lastly of the fourth, *iv*; the third or middle toe, *iii*, being the most constant and important of the five toes. The same law or progress of simplification prevails in the fore-foot or hand. The thumb is the first to disappear, then the little finger, and the middle finger is the most constant, and forms the single-hoofed fore-foot of the horse.

The scapula, 51, in the fore-limb repeats or answers to the ilium, 62, in the hind-limb; the coracoid, 52, to the ischium, 63; the clavicle, 58, to the pubis, 64; the humerus, 53, to the femur, 65; the radius, 55, to the tibia, 66; the ulna, 54, to the fibula, 67; the carpus, 56, repeats the tarsus, 68; and the metacarpus and phalanges of the fore-foot repeat the metatarsus and phalanges of the hind-foot: they are technically called "serial homologues," or "homotypes," and each bone in the carpus can be shown to have its homotype in the tarsus. (See *Archetype of the Vertebrate Skeleton*," p. 167.)

SKELETON OF THE SLOTH.

The transition from the quadrupeds with hoofs to those with claws seems in the present series to be abrupt; but it was made gradual by a group of animals, now extinct, which combined hoofs and claws in the same foot. Some of the outer toes, at least, were stunted and buried in a thick callosity, for the ordinary purpose of walking, whilst the other toes were provided with very long and strong claws for uprooting or tearing off the branches of trees. These singular beasts were of great bulk, and appear to have been peculiar to America. As restored by anatomical science, they have received the names of *Megatherium*, *Megalonyx*, *Mylodon*, &c. They were huge terrestrial sloths; the present remnants of the family consist of very few species enabled by their restricted bulk to climb the trees in quest of their leafy food, and peculiarly organized for arboreal life. The toes, which were modified in their huge predecessors to tread the ground, are reduced to rudiments, or are undeveloped; and those only are retained which support the claws, now rendered by their length and curvature admirable instruments for clinging to the branches. The whole structure of the hind and fore limbs is modified to give full effect to these instruments as movers and suspenders of the body in the bosky retreats for which the sloths are destined; and, in the same degree, the power of the limbs to support and carry the animal along the bare ground is abrogated. Accordingly, when a sloth is placed on level ground, it presents the aspect of the most helpless and crippled of creatures. It is less able to raise its trunk above its limbs than the seal, and can only progress by availing itself of some inequality in the soil offering a holdfast to its claws, and enabling it to drag itself along. But to judge of the creative dispensations towards such an animal by observation of it, or reports of its procedure, under these unnatural circumstances, would be as reasonable as a speculation on the natural powers of a tailor suddenly transferred from his shopboard to the rigging of a ship under way, or of a thorough-bred seaman mounted for the first time on a full-blood horse at Ascot. Rouse the prostrate sloth, and let it hook on to the lower bough of a tree, and the comparative agility with which it mounts to the topmost branches will surprise the spectator. In its native South American woods its agility is still more remarkable, when the trees are agitated by a storm. At that time, the instinct of the sloth teaches it that the migration from tree to tree will be most facilitated. Swinging to and fro, back downwards, as is its habitual position, at the end of a branch just strong enough to support the animal, it takes advantage of the first branch of the adjoining tree that may be swayed by the blast within its reach, and stretching out its fore-limb, it hooks itself on, and at once transfers itself to what is equivalent to a fresh pasture. The story of the sloth voluntarily dropping to the ground, and crawling under pressure of starvation to another tree, is one of the fabulous excrescences of a credulous and gossiping zoology.

In the sloth, accordingly (Fig. 40), the fore-limbs are much elongated, and that less at the expense of the hand than of the arm and forearm. The humerus, 53, is of

Fig. 40.

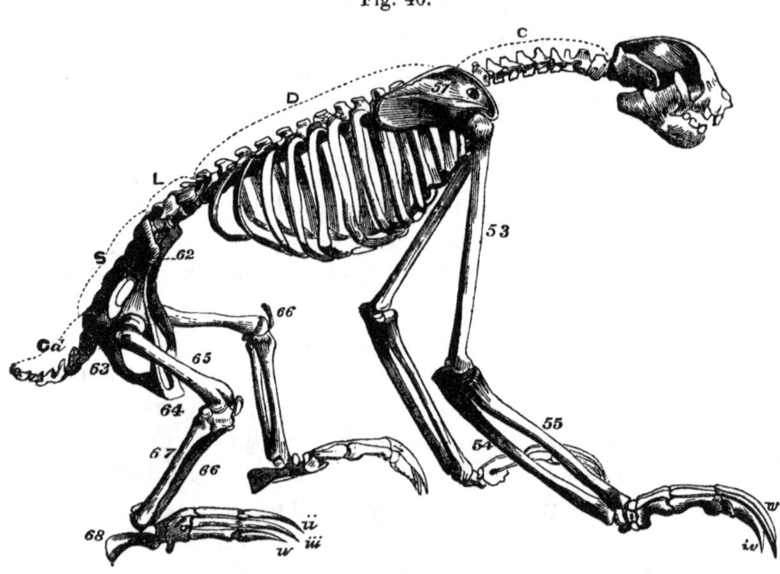

SKELETON OF THE SLOTH.

unwonted length—is slender and straight; the radius, 55, and ulna, 54, are of similar proportions—the former straight, the latter so bent as to leave a wide interosseous space. Now, moreover, these bones, instead of being firmly united as one bone, are so articulated with each other as to permit a reciprocal rotatory movement, chiefly performed, however, by the radius; and since to this bone the carpal segment of the hand is mainly articulated, that prehensile member can be turned prone or supine, as in the human arm and hand. Six bones are preserved in the carpus of three-toed sloth (*Bradypus tridactylus*), answering to those called "lunare," "cuneiforme," "unciforme," and "pisiforme," also to the "scaphoides and trapezium" united, and to the "trapezoides and magnum" united. The scapho-trapezium is characteristic of the sloth-tribe, and is found in the extinct as well as existing species. The articulation of the carpus with the radius, and with the metacarpus, is such as to turn the palm of the long hand inwards, and bring its outer edge to the ground. The three fully-developed metacarpals are confluent at their base, which is also anchylosed to the rudiments of the first and fifth metacarpals; the proximal phalanges of the digits answering to *ii*, *iii*, and *iv*, are confluent with their metacarpals, and those digits appear therefore to have only two joints. The last phalanx is remarkably modified for the attachment of the very long and strong claw.

With regard to the bladebone of the sloth, 51, it is much broader in proportion to its length than in the swift cloven-footed herbivores; the spinous process is unusually short; the acromion is of moderate length, and unexpanded at its extremity; the supraspinal fossa is the broadest, and has a perforation instead of the usual "supraspinal" notch. There is a short clavicular bone attached to the acromion, but not attaining to the sternum.

The iliac bones, 62, repeat the modifications of their homotypes the scapulæ, and are of unusual breadth as compared with those of other quadrupeds; they soon become anchylosed to the broad sacrum, S, the ischia, 63, and pubes, 64, are long and slender, and circumscribe unusually large "thyroid" and "ischial" foramina, the latter being completed by the coalescence of the tuberosities of the ischia with the transverse processes of the last two sacral vertebræ. The head of the femur, 65, has no impression of a ligamentum teres. The patella, 66′, is ossified; there is a fabella behind the external condyle. The tibia, 66, and fibula, 67, are bent in opposite directions, intercepting a very wide interosseous space. The anchylosis of their two extremities, which has been found in older specimens, has not taken place here. The inner malleolus projects backwards and supports a grooved process. The outer malleolus projects downwards, and fits like a pivot into a socket in the astragalus, turning the sole of the foot inwards—a position like that of the hand—best adapted for grasping boughs. The calcaneum, 68, is remarkably long and compressed. The scaphoid, cuboid, and cuneiform bones have become confluent with each other and the metatarsals, of which the first and fifth exist only in rudiment. The other three have likewise coalesced with the proximal phalanges of the toes which they support: these toes answer to the second, third, and fourth, in the human foot.

The short and small head of the sloth is supported on a long and flexible neck presenting the very unusual character in the Mammalian class of nine vertebræ, C—the superadded two, however, appearing to have been impressed from the dorsal series, D, by their short, pointed, and usually movable ribs. The head and mouth can be turned round every part of a branch in quest of the leafy food by this mechanism of the neck. As the trunk is commonly suspended from the limbs with the back downwards, the muscles destined for the movements of the back and support of the head are feebly developed, and the vertebral processes for their attachment are proportionally short. The spines of the neck-vertebræ are of more equal length than in most mammals—that of the dentata being little larger than the rest: the spines gradually subside in the posterior dorsals, and become obsolete in the lumbar vertebræ. The first pair of fully-developed ribs, marking the beginning of the true "dorsal" series of vertebræ, are anchylosed to the breast-bone which consists of eight ossicles. In the two-toed sloth, however, which has twenty-three dorsal vertebræ, there are as many as seventeen subcubical sternal bones in one long row, with their angles truncated for the terminal articulations of the sternal ribs, which are ossified.

The skull of the sloth is chiefly remarkable for the size, shape, and connections of the malar bone, which is freely suspended by its anterior attachment to the maxillary and frontal, and bifurcates behind; one division extending downwards, outside the lower jaw, the other ascending above the free termination of the zygomatic process of the squamosal. The premaxillary bone is single and edentulous, being represented only by its palatal portion completing the maxillary arch, but not sending any processes upwards to the nasals.

The skull in the toothless ant-eater chiefly forms a long, slender, slightly-bent bony sheath for its still longer and more slender tongue, the main instrument for obtaining its insect food. The mouth in the living animal is a small orifice at the end of the tubular muzzle, just big enough to let the vermiform tongue glide easily in and out. The fore-limbs are remarkable for the great size and strength of the claw developed from the middle digit: this is the instrument by which the ant-eater mainly effects the breach in the walls of the termite fortresses, which it habitually besieges in order to prey upon their inhabitants and constructors. As in the sloths, both fore and hind feet have an inclination inwards, whereby the sharp ends of the long claws are prevented from being worn by that constant application to the ground which must have resulted from the ordinary position of the foot. The trunk-vertebræ of the ant-eater are chiefly remarkable for the number of accessory joints by which they are articulated together. This complex structure is also met with in the armadillos, in whieh the anterior zygapophyses of the dorsal vertebræ send processes—the metapophyses (Fig. 2, p. 165), *m*, *m*—upwards, outwards, and forwards, which processes, progressively increasing in the hinder vertebræ, attain, in the lumbar region, a length equal to that of the spinous processes, *ns*, and have the same relation to them, in the support of the osseous carapace, as the "tie-bearers" have to the "king-post" in the architecture of a roof.

SKELETON OF THE MOLE.

The mole is hardly less fitted for the actions of an ordinary land-quadruped than the sloth; but the one is as admirably constructed for subterraneous as the other for arboreal life. The fore-limbs are remarkably short, broad, and massive in the mole, as they are long and slender in the sloth; yet the same osseous elements, similarly disposed, occur in the skeleton of each. The head of the mole is long and cone-shaped; its broad base joins on the trunk without any outward appearance of a neck. The forepart of the trunk, to which the principal muscular masses working the fore-limbs are attached, is

Fig. 41.

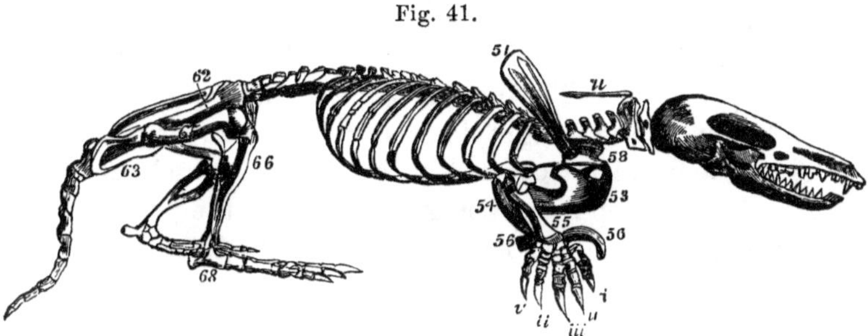

SKELETON OF THE MOLE.

the thickest, and thence the body tapers to the hind-quarters, which are supported by limbs as slender as they are short.

The neck-bones, nevertheless, are not wanting; they even exist in the same number as in the giraffe; the vertebral formula of the mole being—7 cervical, 13 dorsal, 6 lumbar, 5 sacral, and 10 caudal. The spine of the second vertebra or dentata is large, and extended back over the third vertebra: the neural arches of this and the succeeding neck-vertebræ form thin simple arches without spines: the entire vertebræ have been described as mere rings of bone; but the transverse processes of the fourth, fifth, and sixth cervicals are produced forwards and backwards, and overlap each other: in the seventh vertebra those processes are reduced to tubercular diapophyses which are not perforated: the bodies of the vertebræ are depressed and quadrate. The part answering to the nuchal ligament in the giraffe is bony in the mole, *u*.

The first sternal bone, or manubrium, is of unusual length, being much produced forwards, and its under surface downwards in the shape of a deep keel for extending the origin of the pectoral muscles. Seven pairs of ribs directly join the sternum, which consists of four bones, in addition to the manubrium and an ossified ensiform appendage. The neural spines, which are almost obsolete in the first eight dorsals, rapidly gain length in the rest, and are antroverted in the last two dorsal vertebræ. The diapophyses, being developed in the posterior dorsals, determine the nature of the longer homologous processes in the lumbar vertebræ.

The lumbar spines are low, but of considerable antero-posterior extent: the diapophyses are bent forward in the last four vertebræ: a small, detached, wedge-shaped hypapophysis is fixed into the lower interspace of the bodies of the lumbar vertebræ.

The scapula, 51, is very long and narrow, but thick, and almost three-sided: the common rib-shape is resumed in this cranial pleurapophysis, as we have seen in the bird and tortoise. The clavicle, on the other hand, instead of the usual long and slender figure, presents the form of a cube, being very short and broad, articulated firmly to the anteriorly projecting breast-bone, and more loosely with the acromion and head of the humerus.

This bone, 53, would be classified amongst the "flat" bones. It is almost as broad as it is long, especially at its proximal end, which presents two articular surfaces —one for the scapula, the other for the clavicle: the expanse of the bone beyond these surfaces relates to the formation of an adequate extent of attachment for the deltoid, pectoral, and other great burrowing muscles. All the other bones of the fore-limb are as extremely modified for fossorial actions. The olecranon expands transversely at its extremity, and the back part of the ulna is produced into a strong ridge of bone.

The shaft of the radius is divided by a wide interosseous space from the ulna, and the head of the radius is produced into a hook-shaped process like a second "olecranon." The carpal series consists of five bones in each row—the scaphoid being divided in the first, and a sesamoid being added to the second row; moreover, there is a large supplementary sickle shaped bone, extending from the radius to the metacarpal of the pollex, giving increased breadth and a convex margin to the radial side of the very powerful hand, and chiefly completing its adaptation to the act of rapidly displacing the soil. The phalanges of the fingers are short and very strong: the last are bifid at their ends for a firmer attachment of the strong claws. Little more of the hand than these claws, and the digging or scraping edge, projects beyond the sheath of skin enveloping the other joints, and connecting the hand with the trunk.

The common position of the arm-bone is with its distal end most raised. The fore-arm, with the elbow raised, is in the state between pronation and supination, the radial side of the hand being downwards, and the palm directed outwards. The whole limb, in its position and structure, is unequalled in the vertebrate series as a fossorial instrument, and only paralleled by the corresponding limb in the mole-cricket (*Gryllotalpa*) amongst the insect-tribes.

No impediment is offered by the hinder parts of the body or limbs when the thickest part of the animated wedge has worked its way through the soil. The pelvis is remarkably narrow. The ossa innominata have coalesced with the sacrum, but not with each other, the pubic arch remaining open. The bodies of the sacral vertebræ are blended together, and are carinate below; their neural spines have coalesced to form a high ridge. The acetabula look almost directly outwards. The head of the femur has no pit for a round ligament. A fabella is preserved behind the outer condyle. A hamular process is sent off from the head of the tibia and fibula; the lower moieties of the shafts of these bones are blended together. The toes are five in number on the hind-feet as in the fore, but are much more feebly developed. They serve to throw back the loose earth detached by the spade-shaped hands.

SKELETON OF THE BAT.

The form of limb presented by the arm and hand of the bat, offers the most striking contrast to the burrowing trowel of the mole. Viewed in the living animal, it is a thin, widely expanded sheet of membrane, sustained like an umbrella by slender rays, and flapped by means of these up and down in the air, and with such force and rapidity, as, combined with its extensive surface, to react upon the rare element more powerfully than gravitation can attract the weight to which the fore-limbs are attached; consequently, the body is raised aloft, and borne swiftly through the air. The mammal now rivals the bird in its faculty of progressive motion; it flies, and the instruments of its aerial course are called "wings." The whole frame of the bat is in harmony with this faculty, but the mammalian type of skeleton is in nowise departed from.

The vertebral formula of the common bat (*Vespertilio*

murinus) (Fig. 42), is: 7 cervical, 12 dorsal, 7 lumbar, 3 sacral, and 8 caudal vertebræ. The chief characteristics of the skeleton are: the gradual diminution of size of the spinal column from the cervical to the sacral regions; the absence of neural spines in the vertebræ beyond the dentata; a keeled sternum; long and strong, bent clavicles,

Fig. 42.

SKELETON OF THE BAT (*Vespertilio murinus*).

58; broad scapulæ, 51; elongated humeri, 53; more elongated and slender radius, 55; and still longer and more slender metacarpals and phalanges of the four fingers, *ii*, *iii*, *iv*, *v*, which are without claws, the thumb, *i*, being short, and provided with a claw; the pelvis, 62, is small, slender, and open at the pubis, 63; the fibula, 67, is rudimental, like the ulna, 54, in the fore-arm. The common bat has a long and slender stilliform appendage to the heel, 68, which helps to sustain the caudo-femoral membrane. The hind digits are five in number, short, subequal, each provided with a claw; they are the instruments by which the bat suspends itself, head downwards, during its daily summer sleep, and continued winter torpor.

SKELETON OF THE CÁRNIVOROUS MAMMALIA.

The lion may be regarded as the type of a quadruped. The well-adjusted proportions of the head, the trunk, the fore-limbs, and tail concur with their structure to form an animal swift in course, agile in leaps and bounds, terrible in the overpowering force of the blows inflicted by the fore-limbs. The strong, sharp, much-curved, retractile talons, terminating the broad powerful feet, enable the carnivore to seize the prey it has overtaken, and to rend the body it has struck down. The jaws have a proportional strength, and are armed with fangs fitted to pierce, lacerate, and kill.

The carnivorous character of the skull, as exemplified by the sagittal and occipital crests, by the strength and expanse of the zygomatic arches, by the breadth, depth, and shortness of the jaws, by the height of the coronoid processes, and by the depth and extent of the fossæ of the lower jaw for the attachment of the biting muscles, reaches its maximum in the lion. The triangular occipital region is remarkable for the depth and boldness of the sculpturing of its outer surface, indicative of the powerful muscles working the whole skull upon the neck and trunk. The conjoined paroccipitals and mastoids form a broad and thick capsular support for the back part of the acoustic bullæ. The pterygoid processes are imperforate. A well-marked groove extends on each side of the bony palate from the posterior to the anterior palatine foramina. The premaxillaries are comparatively short, and one-half of the lateral border of the nasals directly articulates with the maxillaries. The antorbital foramina are largely indicative of the size of the sensitive nerve supplying the well-developed whiskers. Within the cranium we find that ossification has extended into the membrane dividing the cerebrum from the cerebellum. This bony tentorium extends above the petrosal to the ridge overhanging the Gasserian fossa; the petrosal is short, its apex is neither notched nor perforated; the cerebellar pit is very shallow. The sella-turcica is deep, and well defined by both the anterior and posterior clinoids. The rhinencephalic fossa is relatively larger in the lion than in most carnivora, and is defined by a well-marked angle of the inner table of the skull from the prosencephalic compartment. The olfactory chamber extends backwards both above and below the rhinencephalic fossa; the upper part of the chamber is divided into two sinuses on each side. The superior turbinals extend into the anterior sinus, and below into the presphenoidal sinus. All the bones of the skeleton are remarkable for their whiteness and compact structure.

The vertebral formula of the lion (Fig. 43) is—7 cervical, 13 dorsal, 7 lumbar, 3 sacral, and 23 caudal. The last cervical vertebra has the transverse processes imperforate, being formed only by diapophyses. The eleventh dorsal is that toward which the spines of the other trunk-vertebræ converge, and indicates the centre of motion of the trunk in this bounding quadruped. Eight pairs of ribs directly join the sternum, which consists of eight bones. The clavicles are reduced to clavicular bones, 58, suspended in the flesh. The supraspinal fossa of the scapula is less deep than the infraspinal one, and its border is almost uniformly convex; the acromion is bifid, the recurved point being little larger than the extremity or anterior point. The humerus, 53, is perforated

Fig. 43.

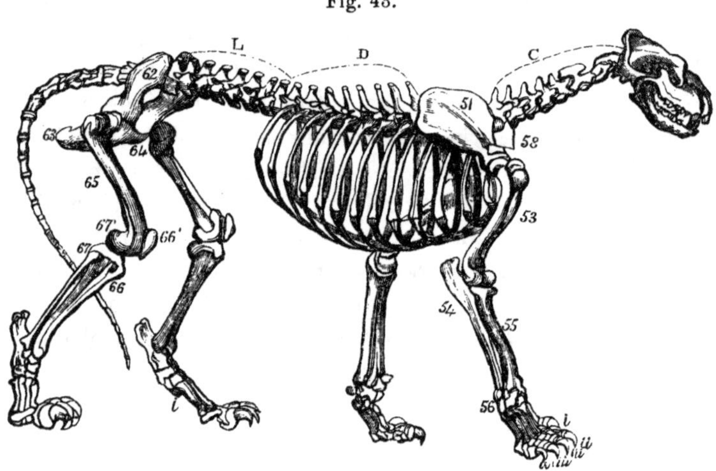

SKELETON OF THE LION (*Felis leo*).

above the inner condyle, but not between the condyles. The radius, 55, and ulna, 54, are so articulated as to permit a free rotation of the fore-paw. The scaphoid and lunar bones are connate. Besides these, the bones of the carpus are the cuneiforme; the pisiforme; the trapezium, which gives an articulation to the ulnar side of the base of the short metacarpus of the pollex; the trapezoides; the magnum, which is the least of the carpal bones; the unciforme, which supports, as usual, the metacarpals of the fourth and fifth digits; and the pisiforme, which projects far backwards, like a small calcaneum: there is also a supplementary oscicle wedged in the interspace between the prominent end of the scapho-lunar bone and the proximal end of the metacarpal of the pollex. The pollex is retained on the fore-foot, and, like the other toes, is terminated by a large, compressed, retractile, ungual phalanx, forming a deep sheath, for the firm attachment of the large curved and sharp-pointed claws.

The pelvis, 62, 63, 64, the femur, 65, the tibia, 66, and fibula, 67, offer no remarkable modifications of structure; the patella, 66, is well ossified, and there is a fabella, 67, behind each condyle of the femur. The tarsal bones are the astragalus; the scaphoides; the calcaneum; the cuboides, which, like the unciforme in the carpus, supports the two outer digits; the cuneiforme externum, which, like the magnum, supports the middle digit; the cuneiforme medium, which, like the trapezoides, supports the second digit; and the cuneiforme internum, which supports the rudiment of the metatarsal of the first or innermost digit.

The last or ungual phalanx, in both fore and hind feet, has a bony sheath at its base for the firmer implantation of the claw; and its joint is at the back part of the proximal end of the phalanx, whereby it can be drawn upwards upon the second phalanx, when the claw becomes concealed in the fold of integument forming the interspace of the digits.

This state of retraction is constantly maintained, except when overcome by an extending force, by means of elastic ligaments. The principal one arises from the outer side and distal extremity of the second phalanx, and is inserted into the superior angle of the last phalanx; a second arises from the outer side and proximal end of the second phalanx, and passes obliquely to be inserted at the inner side of the base of the last phalanx. A third, which arises from the inner side and proximal extremity of the second phalanx, is inserted at the same point as the preceding. The tendon of the flexor profundus perforans is the antagonist of the elastic ligaments. By the action of that muscle, the last phalanx is drawn forwards and downwards, and the claw exposed. In order to produce the full effect of drawing out the claw, a corresponding action of the extensor muscle is necessary, to support and fix the second phalanx; by its ultimate insertion in the terminal phalanx, it serves also to restrain and regulate the actions of the flexor muscle. As the phalanges of the hind-foot are retracted in a different direction to those of the fore-foot, *i. e.* directly upon, and not by the side of the second phalanx, the elastic ligaments are differently disposed, but perform the same main office.

It seems scarcely necessary to allude to the final intention of these beautiful structures, which are, with some slight modifications, common to the genus *Felis.* The claws being thus retracted within folds of the integument, are preserved constantly sharp, and ready for their destined functions, not being blunted and worn away in the ordinary progressive motions of the animal; while at the same time the sole of the foot, being padded, such soft parts only are brought in contact with the ground as conduce to the noiseless tread of the stealthy feline tribe. This highly-developed unguiculate structure, with the dental system and concomitant modifications of the skull, complete the predatory character of the typical *Carnivora.*

SKELETON OF THE KANGAROO.

Australia possesses an indigenous race of herbivorous mammals, created to enjoy existence on its grassy plains. But the climate of this fifth continent, as, from its extent, it has been termed, is subject to droughts of unusual duration; and the parched up grass, ignited by the electric bolt or other cause, often raises a conflagration of fearful extent, and leaves a correspondingly wide-spread blackened desert. To the antelope, and other ruminants of

Fig. 44.

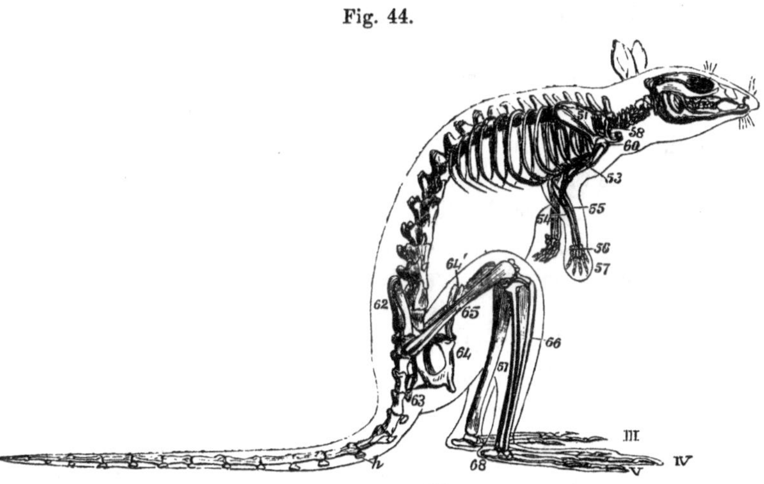

SKELETON OF THE KANGAROO (*Macropus elegans*).

tropical or warmer latitudes, swiftness of limb has been given, which enables them to migrate to river valleys, where the vegetation is preserved from the influence of the dry season. Australia, however, is peculiar for its scanty supply of perennial streams; the torrents of the brief periods of rain are reduced to detached pools in the dry season, and these are parched up in the long droughts, leaving hundreds of miles of the country devoid of surface water. If, then, the parent herbivore could traverse the required distance to quench its thirst, or satisfy its hunger, the tender young would be unable to follow the dam. A modification of the procreative process has accordingly been superinduced, which characterizes the Australian mammals; the young are prematurely brought forth of embryonic size and helplessness, and are transferred to a pouch of inverted skin, concealing the udder; and in this marsupium, as in a well-stored vehicle, they are easily transferred by the parent to any distance to which the climatal conditions may compel her to migrate. The economy of this portable nursery, the requisite manipulation of the suckling young therein suspended from the teat, demand a certain prehensile power of the fore-limbs, a freedom of the digits, with some opposable faculty in them, and the possession of so much sense of touch as would be impossible were the digit to be incased in a hoof; the horny matter is accordingly developed only on the upper surface of the finger-end, and is in the form of a claw. But the unguiculate pentadactyle extremity—though a higher grade of structure in the progress of limbs—is not suited for the exigences of the herbivore, and would have appeared utterly incompatible with an existence dependent on grazing in wild pastures, had we argued from knowledge restricted to the forms and structures of the hoofed herbivores of the Europæo-Asiatic, African, and American continents. How, then, it may be asked, is this difficulty overcome in the case of a grazing animal, necessarily a marsupial, and consequently an unguiculate one? The answer need only be a reference to Fig. 44; the requisite faculty of migration of the parent with the tender offspring is gained by transferring the locomotive power to the hinder pair of limbs extraordinarily developed, and aided by a correspondingly powerful tail; the fore-limbs being restricted in their development to the size requisite for the marsupial offices and other accessory uses.

This is the condition or explanation of the seemingly anomalous form and proportions of the kangaroo—so strange, indeed, that the experienced naturalists, Banks and Solander, may well be excused for surmising they had seen a huge bird when they first caught a glimpse of the kangaroo in the strange land which they, with Cook, discovered.

The rapid course of the kangaroo is by a succession of leaps, in which twenty to thirty yards are cleared at a bound; the herbivore, instead of a swift courser on four pretty equally developed hoofed extremities, is, in Australia, a leaping animal; and the saltatorial modification of the mammalian skeleton is here shown in that of one of the swiftest and most agile of the numerous species of kangaroo, the *Macropus elegans*.

In this kangaroo, 13 vertebræ are dorsal, 6 are lumbar, 2 are sacral, and 28 are caudal, the first fourteen of which have hæmapophyses. These elements coalesce at their distal ends, and form small hæmal arches; they overspan and protect from pressure the great bloodvessels of the tail, the powerful muscular fasciculi of which derive increased surface of attachment from these hæmal arches. The pelvis is long; the strong prismatic ilia, 52, and the ischia, 63, carry out the great flexors and extensors of the thigh to a distance from their point of insertion—the femur, which makes these muscles operate upon that lever at a most advantageous angle; the trunk, borne along in the violent leaps, needs to be unusually firmly bound to the pelvic basis of the chief moving powers. Accordingly, we find a pair of bones, 64′, extending forwards from the pubic symphysis, 64, along the ventral walls, giving increased bony origin to the unusually developed median abdominal muscles attaching the thorax to the pelvis; and these "marsupial bones," as they are called, have accessory functions relating to reproduction in both sexes of the marsupial quadrupeds. The femur, 65, is more than twice the length of the humerus; it is proportionally strong, with well-developed great and small "trochanters," and a "fabella" behind one or both condyles. The patella is unossified. The fibula, 67, is immovably united to the lower half of the tibia. This bone, 66, is of unusual length and strength, and is firmly interlocked below with the trochlear astragalus. The heel-bone sends backwards a long lever-like process for the favorable insertion of the extensors of the foot. This member is of very unusual length. The innermost toe, or hallux, is absent; the second and third toes are extremely slender, inclosed as far as the ungual phalanx in a common fold of integument, and reduced to the function of cleansing the fur. The offices of support and progression are performed by the two outer toes, *iv* and *v*, and principally by the fourth, which is enormously developed, and terminated by a long, strong, three-sided, bayonet-shaped claw; these two toes are supported, as usual, by the os cuboides, which is correspondingly large, whilst the naviculare and the cuneiform bones are proportionally reduced in size. The bones of the fore-limb, though comparatively diminutive, present all the complexities of structure of the unguiculate limb. The clavicle, 58, connects the acromion with the sternum, and affords a fulcrum to the shoulder-joint. The humerus, articulating below with a radius and ulna which can rotate on each other, develops ridges above both inner and outer condyles for the extended origin of the muscles of pronation and supination. The brachial artery pierces the entocondyloid ridge. The carpal bones, answering to the scaphoid and lunar in the human wrist, are here confluent. The digits are five in number, enjoy free, independent movements, ard are each terminated by a sharp-curved claw.

SKELETON OF THE QUADRUMANA.

The sloth is an exclusively arboreal animal; its diet is foliage; it has but to bring its mouth to the leafy food, and the lips and tongue serve to strip it from the branches. The extremities, as we have seen, serve mainly to climb and cling to branches, and occasionally to hook down a tempting twig within reach of the mouth. There is, however, another much more extensive and diversified order of arboreal mammals destined to subsist on the fruits and other more highly developed products of the vegetable kingdom than mere leaves. In the monkeys, baboons, and apes the extremities are endowed with prehensile faculties of a more perfect and varied character than in the sloths; and this additional power is gained by a full development of the digits in normal number, with free and independent movements, which in one of them—the first or innermost—are such as that it can be opposed to the rest, so that objects of various size can be grasped. This modification converts a foot into a hand; and, as the mammals in question have the opposable "thumb" on both fore and hind limbs, they are called "quadrumana," or four-handed. The rest of the limb manifests a corre-

Fig. 46.

Fig. 45.

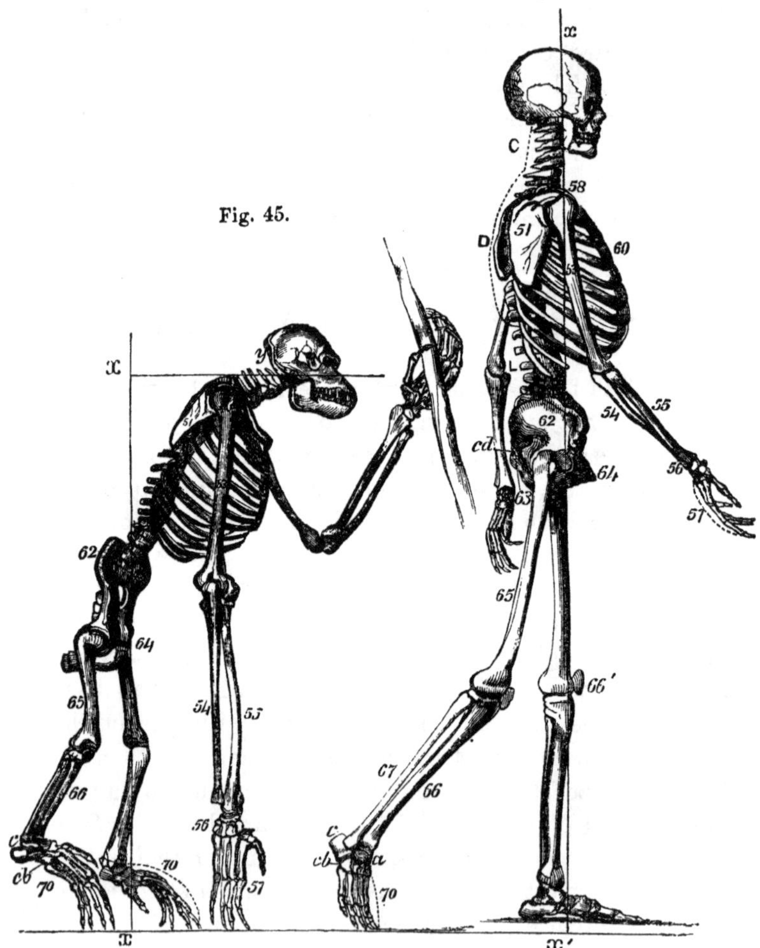

SKELETONS OF ORANG (*Pithecus satyrus*) AND MAN.

sponding complexity or perfection of structure; the trunk is adjusted to accord with the actions of such instruments, and the brain is developed in proportion with the power of executing so great a variety of actions and movements as the four-handed structure gives capacity for.

In the skull of the quadrumana are seen indications of a concomitant perfection of the outer senses; the orbits are entire, and directed forwards, with their outlets almost on the same plane; both eyes can thus be brought to bear upon the same object. The rest of the face, formed by the jaws, now begins to bear a smaller proportion to the progressively expanding cranium. The neck, of moderate length, has its seven vertebræ well developed, with the costal processes large in the fifth and sixth: the dorsal vertebræ, twelve, in the species figured (*Pithecus satyrus*), show, by the convergence of their spines towards the vertical one on the ninth, that this is the centre of movement of the trunk. The lumbar vertebræ are four in number; in the inferior monkeys they are seven, and the anterior ones are firmly interlocked by well-developed anapophyses and metapophyses. The sacrum is still long and narrow. The tail, in some of the lower quadrumana, is of great length, including 30 vertebræ in the red monkey (*Cereopithecus ruber*), in which the anterior ones are complicated by having hæmal arches. The clavicles are entire in all quadrumana. The humerus has its tuberosities and condyloid crests well developed. The radius rotates freely on the ulna. The wrist has nine bones, owing to a division of the scaphoid, besides supplementary sesamoids adding to the force of some of the muscles of the hand; the thumb is proportionally shorter in the fore than in the hind foot. The patella is ossified, and in most baboons and monkeys there is a fabella behind each condyle of the femur. The fibula is entire, and articulated with the tibia at both ends. The tarsus has the same number and relative position of the bones as in man; but the heel-bone is shorter, and the whole foot rather more obliquely articulated upon the leg, the power of grasping being more cared for than that of supporting the body; the innermost toe forms a large and powerful opposable thumb.

There is a well-marked gradation in the quadrumanous series from the ordinary quadrupedal to the more bipedal type. In the lemurs and South American monkeys, the anterior thumb is shorter and much less opposable than the hinder one; in the spider-monkeys it is wanting, and a compensation seems to be given by the remarkable prehensile faculty of the curved and callous extremity of the long tail. This member, in the African and Asiatic monkeys, is not prehensile, but the thumb of the fore-hand is opposable. In the true apes, the tail is wanting, *i.e.* it is reduced to the rudiment called "os coccygis;" but the fore-arms are unusually developed in certain species, hence called "long-armed apes." These can swing themselves rapidly from bough to bough, traversing wide spaces in the aerial leap. The orang (Fig. 45) is also remarkable for the disproportionate length of the arms, but this difference from man becomes less in the chimpanzees. The large species called Gorilla, which of all brutes makes the nearest approach to man, is still strictly "quadrumanous;" the great toe, or "hallux," being a grasping and opposable digit. But the hiatus that divides this highest of the ape tribe from the lowest of the human species, is more strikingly and decisively manifested in the skull (Fig. 50). The common teeth in the male gorilla are developed, as in the male orang, to proportions emulating the tusks of the tiger; they are, however, weapons of combat and defence in these great apes, which are strictly frugivorous. Nevertheless, the muscles that have to work jaws so armed require modifications of the cranium akin to those that characterize the lion, viz: great interparietal, 7, and occipital, 3, cristæ and massive zygomatic arches. The spines of the cervical vertebræ are greatly elongated in relation to the support of such a skull, the facial part of which extends so far in advance of the joint between the head and neck. The chimpanzees, moreover, differ from man in having thirteen pairs of thoracic movable ribs. The long and flat iliac bones, 62, the short femora, 65, so articulated with the leg-bones, 66, as to retain habitually a bent position of the knee, the short calcanea, *c*, and the inward inclination of the sole of the foot, all indicate, in the highest as in the lowest quadrumana, an inaptitude for the erect position, and a compensating gain of climbing power favorable for a life to be spent in trees.

In the osteological structure of man (Fig. 46), the vertebrate archetype is furthest departed from by reason of the extreme modifications required to adjust it to the peculiar posture, locomotion, and endless variety of actions characteristic of the human race.

As there is nothing, short of flight, done by the moving powers of other animals that serpents cannot do by the vertebral column alone, so there is no analogous action or mode of motion that man cannot perform, and mostly better, by his wonderfully developed limbs. The reports of the achievements of our athletes, prize wrestlers, prize pedestrians, funambulists, and the records of the shark-pursuing and shark-slaying amphibious Polynesians, of the equestrian people of the Pampas, of the Alpine chasers of the chamois, and of the scansorial bark-strippers of Aquitaine, concur in testifying to the intensity of those varied powers, when educed by habit and by skilled practice. The perfection of almost all modifications of active and motive structures seems to be attained in the human frame, but it is a perfection due to especial adaptation of the vertebrate type, with a proportional departure from its fundamental pattern. Let us see how this is exemplified in the skeleton of man (Fig. 46), viewing it from the foundation upwards.

In the typical mammalian foot, the digits decrease from the middle to the two extremes of the series of five toes; and in the modifications of this type, as we have traced them through the several gradations (pp. 185, 187, Figs. 35–39), the innermost, *i*, is the first to disappear. In man, it is the seat of excessive development, and receives the name of "hallux," or "great toe;" it retains, however, its characteristic inferior number of phalanges. The tendons of a powerful muscle, which in the orang and chimpanzee are inserted into the three middle toes, are blended in man into one, and this is inserted into the hallux, upon which the force of the muscle now called "flexor longus pollicis" is exclusively concentrated.

The arrangement of other muscles, in subordination to the peculiar development of this toe, makes it the chief fulcrum when the weight of the body is raised by the power acting upon the heel, the whole foot of man exemplifying the lever of the second kind. The strength and backward production of the heel-bone, *c*, relate to the augmentation of the power. The tarsal and metatarsal bones are coadjusted so as to form arches, both lengthwise and across, and receive the superincumbent weight from the tibia on the summit of a bony vault, which has the advantage of a certain elasticity combined with adequate strength. In proportion to the trunk, the pelvic limbs are longer than in any other animal; they even exceed those of the kangaroo, and are peculiar for the superior length of the femur, 65, and for the capacity of this bone to be brought, when the leg is extended, into the same line with the tibia, 66; the fibula, 67, is a distinct bone. The inner condyle of the femur is longer than the outer one, so that the shaft inclines a little outwards to its upper end, and joins a neck longer than in other animals, and set on at a very open angle. The weight of the body, received by the round heads of the thigh-bones, is thus transferred to a broader base, and its support in the upright posture facilitated. The pelvis is modified so as to receive and sustain better the abdominal viscera, and to give increased attachment to the muscles, especially the "glutei," which, comparatively small in other mammals, are in man vastly developed to balance the trunk upon the legs, and reciprocally to move these upon the trunk. The great breadth and anterior concavity of the ilium, 62, are characteristic modifications of this bone in man. The pelvis is more capacious, the tuberosity of the ischium is less prominent, and the symphysis pubis shorter, than in apes. The tail is reduced to three or four stunted vertebræ, anchylosed to form the bone called "os coccygis." The five vertebræ which coalesce to form the sacrum, are of unusual breadth, and the free or "true" vertebræ, that rest on the base of the sacral wedge, gradually decrease in size to the upper part of the chest; all the free vertebræ, divided into five lumbar, twelve dorsal, and seven cervical, are so articulated as to describe three slight and graceful curves, the bend being forward in the loins, backward in the chest, and forward again in the neck. A soft elastic cushion, of "intervertebral" substance, rests between the bodies of the vertebræ. The distribution and libration of the trunk, with the superadded weight of the head and arms, are favored by these gentle curves, and the shock in leaping is broken and diffused by the numerous elastic intervertebral joints. The expansion of the cranium behind, and the shortening of the face in front, give a globe-like form to the skull, which is poised by a pair of condyles, advanced to near the middle of its base upon the cups of the atlas; so that there is but a slight tendency to incline forwards when the balancing action of the muscles ceases, as when the head nods during sleep in an upright posture.

The framework of the upper extremity shows all the perfections that have been superinduced upon it in the mammalian series, viz: a complete clavicle, 58, antibrachial bones, 54, 55, with rotatory movements as well as those of flexion and extension, and the five digits, 57, free and endowed with great extent and variety of movements: of these, the innermost, which is the first to shrink and disappear in the lower mammalia, is in man the strongest, and is modified to form an opposable thumb more powerful and effective than in any of the quadrumana. The scapula, 51, presents an expanded surface of attachment for the muscles which work the arm in its free socket; the humerus, 53, exceeds in length the bones of the fore-arm. The carpal bones, 56, are eight in number, called scaphoides, lunare, cuneiforme, pisiforme, trapezium, trapezoides, magnum, and unciforme: of these the scaphoid and unciforme are compound bones; *i. e.* they consist each of two of the bones of the type-carpus, connate.

In the human skull, viewed in relation to the archetype, as exemplified in the fish and the crocodile, the following extreme modifications have been established. In the occipital segment, the hæmal arch is detached and displaced, as in all vertebrates above fish; its pleurapophysis (scapula, 51) has exchanged the long and slender for the broad and flat form; the hæmapophysis (coracoid, 52) is rudimental, and coalesces with 51. The neurapophyses (exoccipitals, 2) coalesce with the neural spine (superoccipital), and next with the centrum (basioccipital). This afterwards coalesces with the centrum (basisphenoid) of the parietal segment. With this centrum also the neurapophyses, called "alisphenoids," and the centrum of the frontal vertebra, called "presphenoid," become anchylosed. The neural spine (parietal) retains its primitive distinctness, but is enormously expanded, and is bifid, in relation to the vast expansion of the brain in man. The parapophysis (mastoid) becomes confluent with the tympanic, petrosal, and squamosal, and with the pleurapophysis, called "stylohyal," of the hæmal (hyoidian) arch. The hæmapophysis is ligamentous, save at its junction with the hæmal spine, when it forms the ossicle called "lesser cornu of the hyoid bone," the spine itself being the basihyal or body of the hyoid bone. The whole of this inverted arch is much reduced in size, its functions being limited to those of the tongue and larynx, in regard to taste, speech, and deglutition. The neurapophyses (orbitosphenoids) becoming confluent with the centrum (presphenoid) of the frontal vertebra, and the latter coalescing with that of the parietal vertebra, the compound bone called "sphenoid" in anthropotomy results, which combines the centrums and neurapophyses of two cranial vertebræ, together with a diverging appendage (pterygoid) of the maxillary arch.

The knowledge of the essential nature of such a compound bone gives a clue to the phenomena of its development from so many separate points, which final causes could never have satisfactorily afforded. As the centrum, 5, becomes confluent with No. 1, a still more complex whole results, which has accordingly been described as a single bone, under the name of "os spheno-occipital" in some anthropotomies. Such a bone has not fewer than twelve distinct centres of ossification, corresponding with as many distinct bones in the cold-blooded animals that depart less from the vertebrate archetype. The spine of the frontal vertebra (frontal bone) is much expanded and bifid, like the parietal bone; but the two halves more frequently coalesce into a single bone, with which the parapophysis (postfrontal) is connate. The pleurapophysis of the hæmal arch (tympanic bone) is reduced to its function in relation to the organ of hearing, and becomes anchylosed to the petrosal, the squamosal, and the mastoid. The hæmapophysis is modified to form the dentigerous lower jaw, but articulates, as in other mammals, with a diverging appendage (squamosal) of the antecedent hæmal arch, now interposed between it and its proper pleurapophysis; the two hæmapophyses, moreover, become confluent at their distal ends, forming the symphysis mandibulæ.

The centrum of the first or nasal vertebra, like that of the last vertebra in birds, is shaped like a ploughshare, and is called "vomer;" the neurapophyses have been subject to similar compression, and are reduced to a pair of vertical plates, which coalesce together, and with parts of the olfactory capsules (upper and middle turbinals) forming the compound bone called "æthmoid;" of which the neurapophyses (prefrontals) form the "lamina perpendicularis" in human anatomy. The prefrontals assume this confluence and concealed position even in some fishes—*xiphias*, *e. g.*—and repeat the character in all mammalia and in most birds; but they become partially exposed in the ostrich and the batrachia. The spine of the nasal vertebra (nasal bones) is usually bifid, like those of the two succeeding segments; but it is much less expanded. The hæmal, called "maxillary" arch, is formed by the pleurapophyses (palatines) and by the hæmapophyses (maxillaries), with which the halves of the bifid hæmal spine (premaxillaries) are partly connate, and become completely confluent. Each moiety, or premaxillary, is reduced to the size required for the lodgement of two vertical incisors: as the canines in man do not exceed the adjoining teeth in length, and the premolars are reduced to two in number, the alveolar extent of the maxillary is short, and the whole upper jaw is very slightly prominent.

Of the diverging appendages of the maxillary arch, the more constant one, called "pterygoid," articulates with the palatine, but coalesces with the sphenoid; the second pair, formed by the malar, 26, and squamosal, 27, has been subject to a greater degree of modification; it still performs the function assigned to it in lizards and birds, where it has its typical ray-like figure, of connecting the maxillary with the tympanic; but the second division of the appendage (squamosal) which began to expand in the lower mammalia, and to strengthen, without actually forming part, of the walls of the brain-case, now attains its maximum of development, and forms an integral constituent of the cranial parietes, filling up a large cavity between the neural arches of the occipital and parietal segments. It coalesces, moreover, with the tympanic, mastoid, and petrosal, and forms, with the subsequently anchylosed stylohyal, a compound bone called "temporal" in human anatomy. The key to the complex beginning of this "cranial" bone is again given by the discovery of the general pattern on which the skulls of the vertebrate animals have been constructed. In relation to that pattern, or to the archetype vertebrate skeleton, the human temporal bone includes two pleurapophyses, 38 and 28, a parapophysis, 8, part of a diverging appendage, 27, and a sense-capsule, 16.

The departure from the archetype, which we observe in the human skull, is most conspicuous in the neural spines of the three chief segments, which, archetypally, may be regarded as deformities by excess of growth to fulfil a particular use, dependent on the maximization of the brain; the deviation is again marked by arrest of growth or suppression of parts, as *e. g.* in certain parapophyses, and in the hæmal arch of the parietal segment; it is most frequently exemplified in the coalescence of parts primarily and archetypally distinct; and it is finally manifested by the dislocation of a part, viz: the hæmal arch of the occipital segment—the diverging rays of which have become the seat of that marvellous development which has resulted in the formation of the osseous basis of the human hand and arm. With the above explanation the structure of the human skull can be intelligibly comprehended, and not merely empirically understood, as through the absolute descriptions penned in reference to material and utilitarian requirements, and without reference to the great scale of vertebrate structures, of which man is the summit.

The fruit of a series of comparisons, extended over all the vertebrate kingdom, being the recognition of the archetype governing the structure of the vertebrate skeleton, the expression of such knowledge has necessitated the use of general terms, such as "vertebra," for the segments of the skeleton, "neurapophyses," for a constant element of such segment, and the like "general names" for other elements. When any of these elements are modified for special functions, then also a special name for it becomes a convenience, as when a "pleurapophysis" becomes a jaw or blade-bone, &c., a "diverging appendage" an arm or a leg. Deep thinking anatomists have heretofore caught glimpses of these higher, or more general, relations of the vertebral elements, when much modified or specialized, as *e. g.* in the head, and have tried to give expression to the inchoate notion, as when Spix called the "maxillary arch" the "arm of the head." These glimpses of a great truth were, however, ill received; and Cuvier alluded to them, with ill-disguised contempt, as being unintelligible and mystical jargon, in his great work on Fossil Animals (1825). But the error or obscurity lay rather in the mode of stating the relationship of certain bones of the head to those of the trunk, than in the relationship itself: in the endeavor, *e. g.* to express the relation by special instead of general terms. Even in 1845, the learned and liberal-minded editor of Baron Cuvier's last course of lectures, M. de Saint Agy, commenting upon the osteological essays of Spix and Oken, remarks: "For my part, an 'upper-jaw' is an 'upper-jaw,' and an 'arm' is an 'arm.' One must not seek to originate an osteology out of a system of metaphysics."[1] But a jaw is not the less a jaw because it is a "hæmapophysis,"

[1] "Pour moi, une mâchoire superieure est une mâchoire superieure, et un bras est un bras. Il ne faut pas chercher à faire sortir l'osteologie d'un système de metaphysique."

nor is an arm the less an arm because it is a "diverging appendage." In the same spirit a critic might write: "Newton calls this earth a 'planet,' and the moon a 'satellite;' for me the earth is an earth, and the moon is a moon. One must not strive to make an ouranology out of a system of metaphysics." After the first recognition of a thing, one may seek to penetrate, and succeed in knowing, its essential nature, and yet keep within the bounds of nature.

In no class of vertebrate animals is the progressive superiority of the cranium over the face marked by such distinct stages as in the mammalia. Various methods of determining these proportions have been proposed; but the only satisfactory one is by comparing vertical sections of the skull, as in the series figured in the cuts 47—52.

In the cold-blooded ferocious crocodile (Fig. 47), the

Fig. 47.

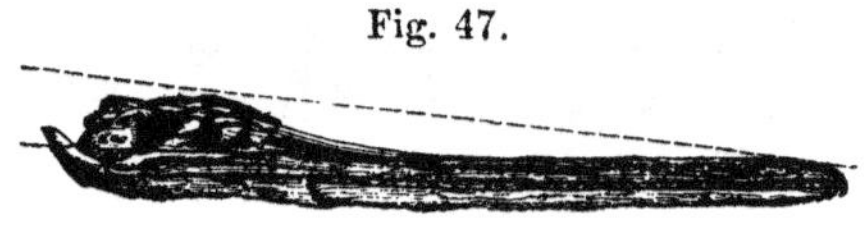

CROCODILE.

cavity for the brain, in a skull three feet long, will scarcely contain a man's thumb. Almost all the skull is made up of the instruments for gratifying an insatiable

Fig. 48.

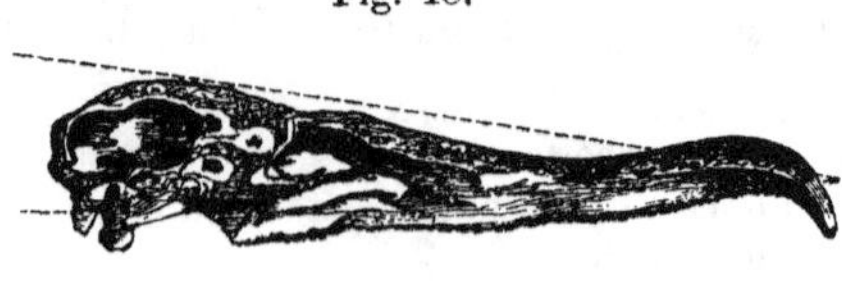

ALBATROSS.

propensity to slay and devour: it is the material symbol of the lowest animal passion.

In the bird (Fig. 48), the brain-case has expanded vertically and laterally, but is confined to the back part of the skull. In the small singing birds, with shorter beaks, the proportion of the cranial cavity becomes much greater.

Fig. 49.

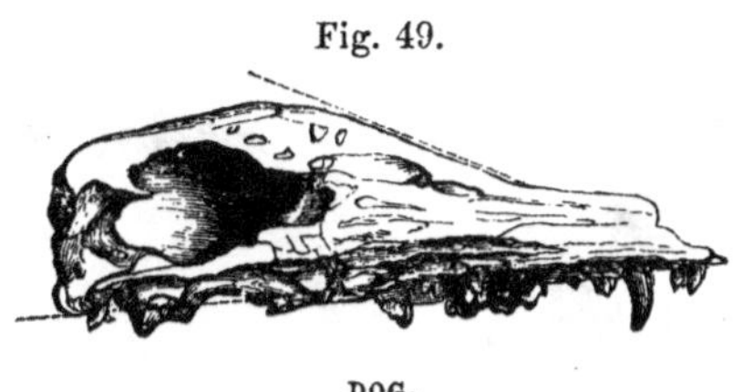

DOG.

In the dog (Fig. 49), the brain-case, with more capacity, begins to advance further forward. In the chimpanzee

Fig. 50.

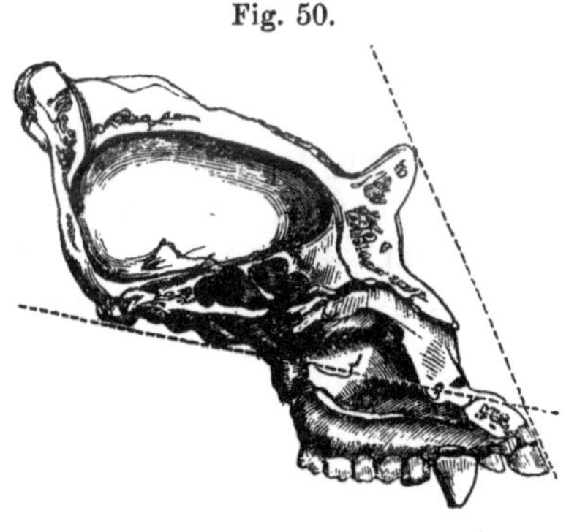

CHIMPANZEE.

(Fig. 50), the capacities or area of the cranium and face are about equal. In man, the cranial area vastly surpasses that of the face.

A difference in this respect is noticeable between the savage (Fig. 51) and civilized (Fig. 52) races of mankind; but it is immaterial as compared with the contrast in this respect presented by the lowest form of the human head (Fig. 51) and the highest of the brute species (Fig. 50).

Fig. 51. Fig 52.

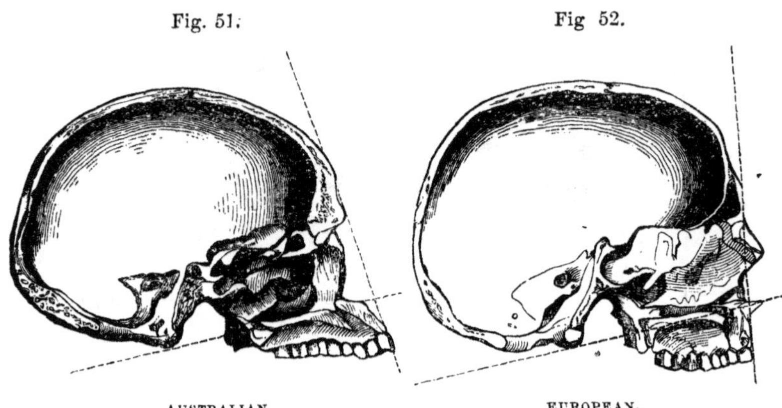

AUSTRALIAN. EUROPEAN.

Such as it is, however, the more contracted cranium is commonly accompanied by more produced premaxillaries and thicker walls of the cranial cavity, as is exemplified in the negro or Papuan skull.

If a line be drawn from the occipital condyle along the floor of the nostrils, and be intersected by a second touching the most prominent parts of the forehead and upper jaw, the intercepted angle gives, in a general way, the proportions of the cranial cavity and the grade of intelligence; it is called the facial angle. In the dog, this angle is 20°; in the great chimpanzee, or gorilla, it is 40°, but the prominent superorbital ridge occasions some exaggeration; in the Australian it is 85°; in the European it is 95°. The ancient Greek artists adopted, in their beau ideal of the beautiful and intellectual, an angle of 100.

CONCLUDING REMARKS.

A retrospect of the varied forms and proportions of the skeletons of animals, whether modified for aquatic, aerial, or terrestrial life, will show that whilst they were perfectly and beautifully adapted to the sphere of life and exigences of the species, they adhered with remarkable constancy to that general pattern or archetype which was first manifested on this planet, as Geology teaches, in the class of fishes, and which has not been departed from even in the most extremely modified skeleton of the last and highest form which Creative Wisdom has been pleased to place upon this earth.

It is no mere transcendental dream, but true knowledge and legitimate fruit of inductive research, that clear insight into the essential nature of each element of the bony framework, which is acquired by tracing them step by step, as *e. g.* from the unbranched pectoral ray of the lepidosiren to the equally small and slender but bifid pectoral ray of the amphiume, thence to the similar but trifid ray of the proteus, and through the progressively superadded structures and perfections of the limbs in higher reptiles and in mammals. If the special homology of each part of the diverging appendage and its supporting arch are recognizable from man to the fish, we cannot close the mind's eye to the evidences of that higher law of archetypal conformity on which the very power of tracing the lower and more special correspondences depend.

Buffon has well remarked, in the Introduction to his great work on Natural History: "It is only by comparing that we can judge, and our knowledge turns entirely on the relations that things bear to those which resemble them and to those which differ from them; so, if there were no animals, the nature of man would be far more incomprehensible than it is."

And if this be true, as to man's general nature and powers, it is equally so with regard to his anatomical structure.

In the same spirit our philosophic poet felt that—

> "'Tis the sublime of man,
> Our noontide majesty, to know ourselves
> Part and proportions of a wondrous whole."—COLERIDGE.

Vertebrated animals of progressively higher grades of structure have existed at successive periods of time on this planet, and they were constructed on a common plan with those that still exist.

Some have concluded, therefore, that the characters of a species became modified in successive generations, and that it was transmuted into a higher species; a reptile, *e. g.* into a mammal; an ape, into a negro. Let us consider, therefore, the import and value of the osteological differences between the gorilla—the highest of all apes—and man, in reference to this "transmutation hypothesis." The skeleton of an animal may be modified to a certain extent by the action of the muscles. By the development of the processes, ridges, and crests, the anatomist judges of the muscular power of the individual to whom a skeleton under comparison has appertained. A very striking difference from the form of the human cranium results from the development of certain crests and ridges for the attachment of muscles, in the great apes; but none of the more important differences, on which the naturalist relies for the determination of the genus and species of the orangs and chimpanzees, have such an origin or dependent relation. The great superorbital ridge, *e. g.* against which the facial line rests in Fig. 50, is not the consequence of muscular action or development: it is characteristic of the genus *Troglodytes* from the time of birth; and we have no grounds for believing it to be a character to be gained or lost through the operations of external causes, inducing particular habits through successive generations of a species.

No known cause of change productive of varieties of mammalian species could operate in altering the size, shape, or connections of the prominent premaxillary bones, which so remarkably distinguish the great *Troglodytes gorilla* from the lowest races of mankind. There is not, in fact, any other character than that founded upon the development of bone for the attachment of muscles, which is known to be subject to change through the operation and influence of external causes. Nine-tenths of the differences which have been cited (see the *Transactions of the Zoological Society*, vol. iii. p. 413), as distinguishing the great chimpanzee from the human species, must stand in contravention of the hypothesis of transmutation and progressive development, until the acceptors of that hypothesis are enabled to adduce the facts demonstrative of the conditions of the modifiability of such characters. Moreover, as the generic forms of the ape tribe approach the human type, they are represented by fewer species. The unity of the human species is demonstrated by the constancy of those osteological and dental peculiarities which are seen to be most characteristic of the *bimana* in contradistinction from the *quadrumana*.

Man is the sole species of his genus (*homo*)—the sole

representative of his order (*bimana*); he has no nearer physical relations with the brute kind than those that belong to the characters which link together the unguiculate division of the mammalian class.

Of the nature of the creative acts by which the successive races of animals were called into being, we are ignorant. But this we know, that as the evidence of unity of plan testifies to the oneness of the Creator, so the modifications of the plan for different modes of existence illustrate the beneficence of the Designer. Those structures, moreover, which are at present incomprehensible as adaptations to a special end, are made comprehensible on a higher principle, and a final purpose is gained in relation to human intelligence; for in the instances where the analogy of humanly invented machines fails to explain the structure of a divinely created organ, such organ does not exist in vain if its truer comprehension, in relation to the Divine idea, or prime Exemplar, lead rational beings to a better conception of their own origin and Creator.

ON THE PRINCIPAL

FORMS AND STRUCTURES OF THE TEETH.

At the commencement of the Treatise on the Principal Forms of the Skeleton, it was stated that "tooth," like "bone," was the result of the combination of certain earthy salts with a pre-existing cellular basis of animal matter. The salts, as shown in a subjoined table, are nearly the same as those in bone, but enter in a larger proportion into the composition of tooth, and render it a harder body. So composed, teeth are peculiar to the back-boned (vertebrate) animals, and are attached to parts of the mouth, commonly to the jaws. They present many varieties as to number, size, form, structure, position, and mode of attachment, but are principally adapted for seizing, tearing, dividing, pounding, or grinding the food. In some species they are modified to serve as formidable weapons of offence and defence; in others, as aids in locomotion, means of anchorage, instruments for uprooting or cutting down trees, or for transport and working of building materials. They are characteristic of age and sex; and in man they have secondary relations subservient to beauty and to speech.

Teeth are always intimately related to the food and

20

habits of the animal, and are, therefore, highly interesting to the physiologist. They form, for the same reason, important guides to the naturalist in the classification of animals; and their value, as zoological characters, is enhanced by the facility with which, from their position, they can be examined in living or recent animals; whilst the durability of their tissues renders them not less available to the palæonthologist in the determination of the nature and affinities of extinct species, of whose organization they are often the sole remains discoverable in the deposits of former periods of the earth's history.

Fig. 53.

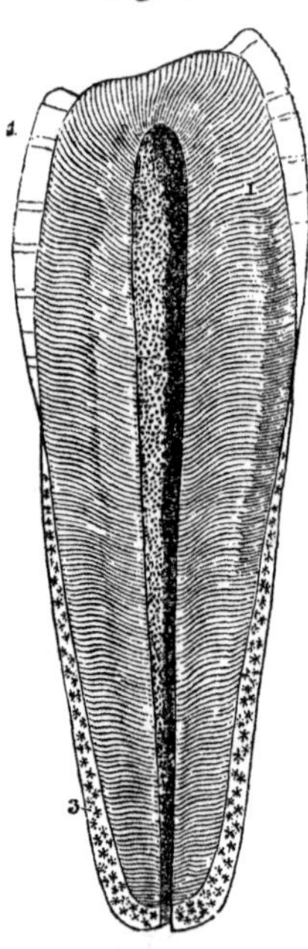

SECTION OF HUMAN INCISOR TOOTH (magnified).

The substance of teeth is not so uniform as in bone, but consists commonly of two or more tissues, characterized by the proportions of their earthy and animal constituents, and by the size, form, and direction of the cavities in the animal basis which contain the earth, the fluid, or the vascular pulp.

The tissue which forms the body of the tooth is called "dentine" (*dentinum*, Lat.), Fig. 53, 1.

The tissue which forms the outer crust of the tooth is called the "cement" (*cæmentum*, and *crusta petrosa*, Lat.), *ib.* 3.

The third tissue, when present, is situated between the dentine and cement, and is called "enamel" (*encaustum*, or *adamas*, Lat.), *ib.* 4.

"Dentine" consists of an organized animal basis disposed in the form of extremely minute tubes and cells, and of earthy particles; these particles have a twofold arrangement, being either blended with the animal matter of the interspaces and parietes of the tubes and cells, or contained in a minutely granular state in their cavities. The density of the dentine arises principally from the proportion of earth in the first of these states of combination. The tubes and cells contain, besides the granular earth, a colorless fluid, probably transuded "plasma," or "liquor sanguinis," and thus relate not only to the mechanical conditions of the tooth, but to the vitality and nutrition of the dentine.

In hard or true dentine, the tubes called "dentinal tubes" diverge from the hollow of the tooth, called "pulp-cavity," and proceed with a slightly wavy course at right angles, or nearly so, to the outer surface. The hard substance of the tooth is thus arranged in hollow columns, perpendicular to the plane of pressure, and a certain elasticity results from their curves: they are upright where the grinding surface of the crown receives the appulse of the opposing tooth, and are horizontal where they have to resist the pressure of contiguous teeth. In Fig. 54, a highly magnified view is given of a small portion of human dentine, showing the tubuli, in the intertubular substance, with the traces of the primitive cellular constitution of that substance. For the mode in which the nucleated cells of the primary basis of the tooth, called "tooth-pulp," are converted into dentine, reference may be made to the author's "Odontography" (Introd., plates 1 and 2). The tubuli, besides fulfilling the mechanical ends above stated, receive the plasma transuded from the remains of the vascular pulp, which circulates, by anastomosing branches of the tubuli and by the plasmatic cells of the intertubular substance, through the dentine, maintaining a sufficient though languid vitality of the tissue. The delicate nerve-branches on the pulp's surface, some minute production

Fig. 54.

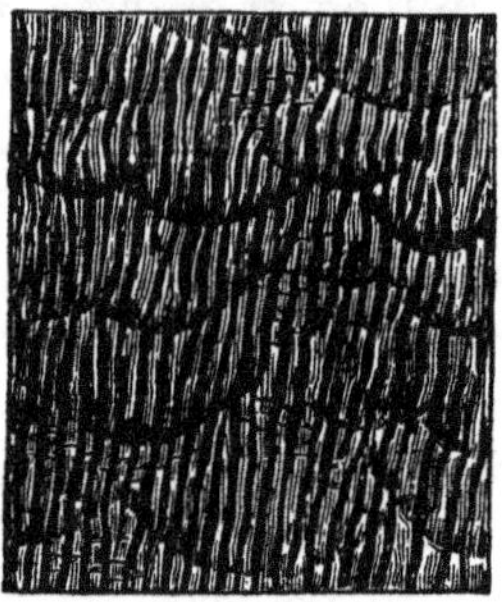

SECTION OF HUMAN TOOTH (highly magnified).

of which may penetrate the tubuli, convey sensations of impressions affecting the dentine—sensations of which every one has experienced the acuteness when decay has affected the dentine, or when mechanical or chemical stimuli have "set the tooth on edge;" but true "dentine" has no canals large enough to admit capillary vessels with the red particles of blood.

The first modification of dentine is that in which capillary tracts of the primitive vascular pulp remain uncalcified, and permanently carry red blood into the substance of the tissue. These so-called "vascular canals" present various dispositions in the dentine which they modify, and which modification is called "vaso-dentine." It is often combined with true dentine in the same tooth, *e. g.*, in the large incisors of certain rodents, the tusks of the elephant, the molars of the extinct iguanodon.

A second modification of the fundamental tissue of the tooth is where the cellular basis is arranged in concentric layers around the vascular canals, and contains "radiated cells" like those of the osseous tissue; it is called "osteo-dentine." The transition from dentine to vaso-dentine, and from this to osteo-dentine, is gradual, and the resemblance of osteo-dentine to true bone is very close.

"Cement" always closely corresponds in texture with the osseous tissue of the same animal; and wherever it occurs of sufficient thickness, as upon the teeth of the horse, sloth, or ruminant, it is also traversed, like bone, by vascular canals. In reptiles and mammals, in which the animal basis of the bones of the skeleton is excavated by minute radiated cells, these are likewise present, of similar size and form, in the cement, and are its chief characteristic as a constituent of the tooth. The relative density of the dentine and cement varies according to the proportion of the earthy material, and chiefly of that part which is combined with the animal matter in the walls of the cavities, as compared with the size and number of the cavities themselves. In the complex grinders of the elephant, the masked boar, and the capybara, the cement, which forms nearly half the mass of the tooth, wears down sooner than the dentine.

The "enamel" is the hardest constituent of a tooth, and consequently the hardest of animal tissues; but it consists, like the other dental substances, of earthy matter, arranged by organic forces in an animal matrix. Here, however, the earth is mainly contained in the canals of the animal membrane, and in mammals and reptiles, completely fills those canals, which are comparatively wide, whilst their parietes are of extreme tenuity.

CHEMICAL COMPOSITION OF TEETH.[1]

	MAN.		LION.		OX.			CROCODILE.		PIKE.
	Dentine.	Enamel.	Dentine.	Enamel.	Dentine.	Enamel.	Cement.	Dentine.	Cement.	Large teeth of lower jaw.
Phosphate of lime with a trace of fluate of lime	66.72	89.82	60.03	83.33	59.57	81.86	58.73	53.69	53.39	63.98
Carbonate of lime	3.36	4.37	3.00	2.94	7.00	9.33	7.22	6.30	6.29	2.54
Phosphate of magnesia	1.08	1.34	4.21	3.70	0.99	1.20	0.99	10.22	9.99	0.73
Salts	0.83	0.88	0.77	0.64	0.91	0.93	0.82	1.34	1.42	0.97
Chondrine	27.61	3.39	31.57	9.39	30.71	6.66	31.31	27.66	28.15	30.60
Fat	0.40	0.20	0.42	trace	0.82	0.02	0.93	0.79	0.76	1.18
	100.00	100.00	100.00	100.00	100.00	100.00	100.00	100.00	100.00	100.00

The examples are extremely few, and, as far as I know, are peculiar to the class *Pisces*, of calcified teeth, which consist of a single tissue, and this is always a modification of dentine. The large pharyngeal teeth of the wrasse (*Labrus*) consist of a very hard kind of dentine.

The next stage of complexity is where a portion of the dentine is modified by vascular canals. Teeth, thus composed of dentine and vasodentine, are very common in fishes.

The hard dentine is always external, and holds the place and performs the office of enamel in the teeth of higher animals. The grinding teeth of the dugong, and the conical teeth of the great sperm whale, are examples of teeth composed of dentine and cement, the latter tissue forming a thick external layer.

In the teeth of the sloth, and its great extinct conge-

[1] Selected from the analytic tables given in the author's "Odontography," 4to. vol. i. pp. lxii., lxiv. (1840.)

ner, the megatherium, the hard dentine is reduced to a thin layer, and the chief bulk of the tooth is made up of a central body of vaso-dentine, and a thick external crust of cement. The hard dentine is, of course, the firmest tissue of a tooth so composed, and forms the crest of the transverse ridges of the grinding surface, like the enamel plates in the elephant's grinder.

The human teeth, and those of the carnivorous mammals, appear at first sight to be composed of dentine and enamel only; but their crowns are originally, and their fangs are always, covered by a thin coat of cement. There is also commonly a small central tract of osteo-dentine in old teeth.

Fig. 55.

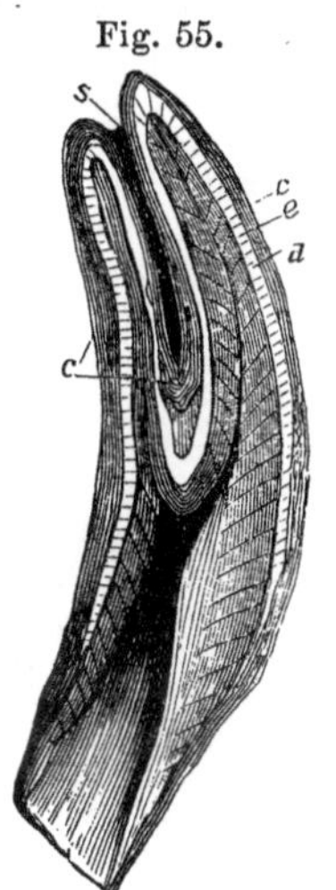

SECTION OF HORSE'S INCISOR.

The teeth, called compound or complex, in *mammalia*, differ as regards their composition from the preceding only by the different proportion and disposition of the constituent tissues. Fig. 55 is a longitudinal section of the incisor of a horse; *d* is the dentine, *e* the enamel, and *c* the cement, a layer of which is reflected into the deep central depression of the crown; *s* indicates the colored mass of tartar and particles of food which fills up the cavity, forming the "mark" of the horse-dealer.

A very complex tooth may be formed out of two tissues by the way in which these may be interblended, as the result of an original complex disposition of the constituents of the dental matrix.

Certain fishes, and a singular family of gigantic extinct

batrachians, which I have called "Labyrinthodonts,"[1] exhibit, as the name implies, a remarkable instance of this kind of complexity. The tooth appears to be of the simple conical kind, with the exterior surface merely striated longitudinally; but, on making a transverse section, as in Fig. 56, each streak is a fissure into which the very thin external layer of cement, *c*, is reflected into the body of the tooth, following the sinuous wavings of the lobes of dentine, *d*, which diverge from the central pulp cavity, *a*. The inflected fold of cement, *c*, runs straight

Fig. 56.

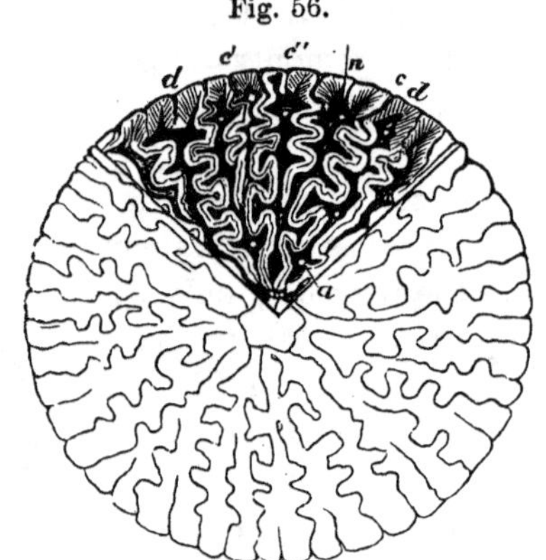

TRANSVERSE SECTION OF TOOTH OF LABYRINTHODON.

for about half a line, and then becomes wavy, the waves rapidly increasing in breadth as they recede from the periphery of the tooth; the first two, three, or four undulations are simple, then their contour itself becomes broken by smaller or secondary waves; these become stronger as the fold approaches the centre of the tooth,

[1] "Proceedings of the Geological Society," Jan. 20, 1841, p. 257.

when it increases in thickness, and finally terminates by a slight dilatation or loop close to the pulp-cavity, from which the free margin of the inflected fold of cement is separated by an extremely thin layer of dentine. The number of the inflected converging folds of dentine is about fifty at the middle of the crown of the tooth, but is greater at the base. All the inflected folds of cement, at the base of the tooth, have the same complicated disposition with increased extent; but, as they approach their termination towards the upper part of the tooth, they also gradually diminish in breadth, and consequently penetrate to a less distance into the substance of the tooth. Hence, in such a section as is delineated (Fig. 56), it will be observed that some of the convoluted folds, as those marked *cc*, extend near to the centre of the tooth; others, as those marked *c'*, reach only about half-way to the centre; and those folds, *c''*, which, to use a geological expression, are "cropping out," penetrate to a very short distance into the dentine, and resemble, in their extent and simplicity, the converging folds of cement in the fangs of the tooth of the ichthyosaurus.

The disposition of the dentine, *d*, is still more complicated than that of the cement. It consists of a slender, central, conical column, excavated, by a conical pulp-cavity, *a*, for a certain distance from the base of the tooth; and this column sends from its circumference, radiating outwards, a series of vertical plates, which divide into two, once or twice, before they terminate at the periphery of the tooth. Each of these diverging and dichotomizing plates gives off, throughout its course, smaller processes, which stand at right angles, or nearly so, to the main plate. They are generally opposite, but sometimes alternate; many of the secondary plates or processes, which are given off near the centre of the tooth, also divide into two before they terminate, as at *n*; and their contour is seen, in the transverse section, to partake of all the undulations of the folds of cement which invest them, and divide the dentinal plates and processes from each other.

Another kind of complication is produced by an aggregation of many simple teeth into a single mass.

The examples of these truly compound teeth are most common in the class of fishes; but the illustration here selected is from the mammalian class. Each tooth of the Cape ant-eater (*Orycteropus*), presents a simple form, is deeply set in the jaw, but without dividing into fangs; its broad and flat base is porous, like the section of a common cane. The canals to which these pores lead, contain processes of a vascular pulp, and are the centres of radiation of as many independent series of dentinal tubules. Each tooth, in fact, consists of a congeries of long and slender prismatic denticles of dentine, which are cemented together by their ossified capsules, the columnar denticles slightly decreasing in diameter, and occasionally bifurcating as they approach the grinding surface of the tooth. Fig. 57 gives a magnified view of a portion of the transverse section of the fourth molar, showing *c*, the cement; *d*, the dentine; and *p*, the pulp-cavity of the denticles.

Fig. 57.

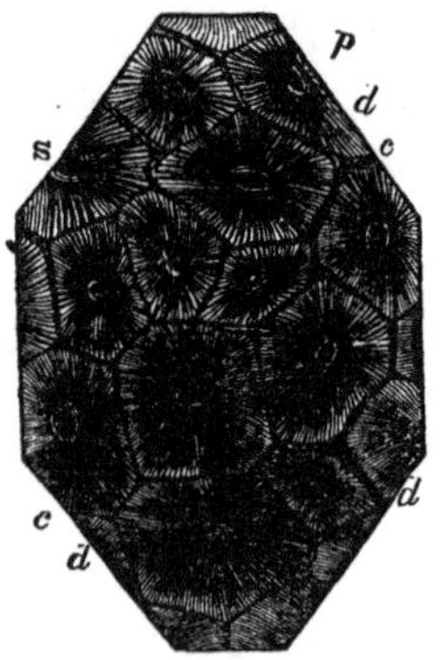

TRANSVERSE SECTION OF PART OF TOOTH OF *Orycteropus* (magnified).

In the elephant, the denticles of the compound molars are in the form of plates, vertical to the grinding surface and transverse to the long diameter of the tooth. When the tooth is bisected vertically and lengthwise, the three substances, *d*, dentine, *e*, enamel, and *c*, cement, are seen interblended, as in Fig. 58, in which *p* is the common pulp-cavity, and *r* one of the roots of this complex tooth.

Such are some of the prominent features of a field of

Fig. 58.

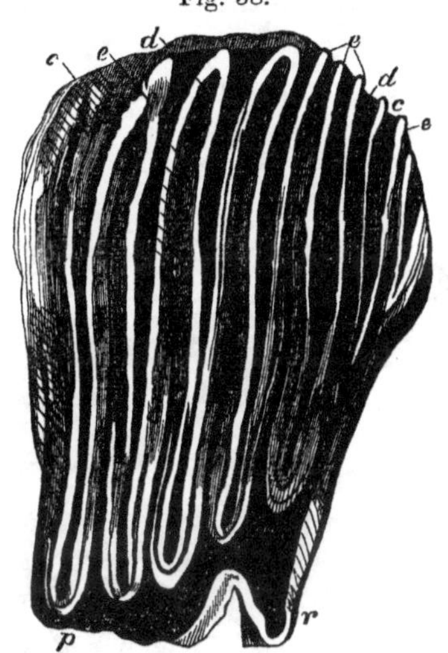

LONGITUDINAL SECTION OF PART OF GRINDER OF ELEPHANT.

observation which Comparative Anatomy opens out to our view—such the varied nature, and such the gradation of complexity of the dental tissues, which, up to December, 1839, continued, notwithstanding successive approximations to the truth, to be described, in systematic works, as a "phaneros," or "a dead part or próduct, exhaled from the surface of a formative bulb!"[1]

[1] See the Fasciculus of M. de Blainville's great work, "Ostéographie et Odontographie d'Animaux Vertébrés," which he submitted to the

DENTAL SYSTEM OF FISHES.

The teeth of fishes, whether we study them in regard to their number, form, substance, structure, situation, or mode of attachment, offer a greater and more striking series of varieties than do those of any other class of animals.

As to number, they range from zero to countless quan-

Fig. 59.

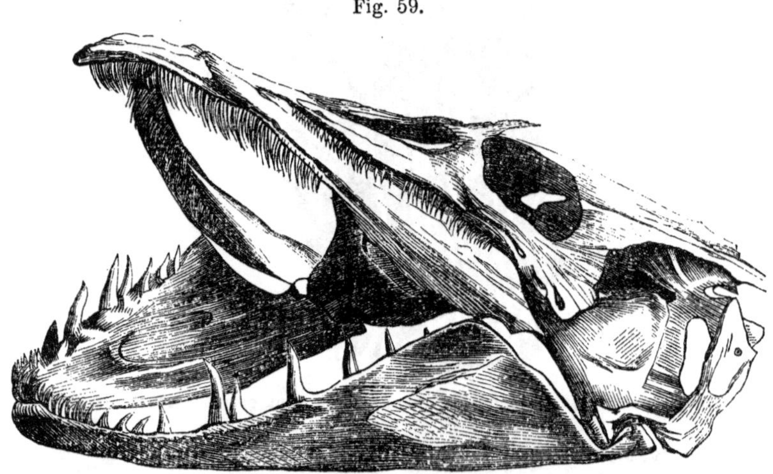

SKULL OF THE PIKE, SHOWING THE TEETH.

tities. The lancelet, the ammocete, the sturgeon, the paddle-fish, and the whole order of *lephobranchii*, are eden-

Academy of Sciences of the Institute of France on the same day, December 16, 1839, on which I communicated, on the occasion of my election as corresponding member of that body, my "Theory of the development of dentine by centripetal calcification and conversion of the cells of the pulp."

tulous. The myxinoids have a single pointed tooth on the roof of the mouth, and two serrated dental plates on the tongue. The tench has a single grinding tooth on the occiput, opposed to two dentigerous pharyngeal jaws below. In the lepidosiren, a single maxillary dental plate is opposed to a single mandibular one, and there are two small denticles on the nasal bone. In the extinct sharks with crushing teeth, called *ceratodus* and *tenodus*, the jaws were armed with four teeth, two above and two below. In the *himæræ*, two mandibular teeth are opposed to four maxillary teeth. From this low point the number in different fishes is progressively multiplied, until, in the pike (Fig. 59), the siluroids, and many other fishes, the mouth becomes crowded with countless teeth.

With respect to form, it may be premised that, as organized beings withdraw themselves more and more, in their ascent in the scale of life, from the influence of the general polarizing forces, so their parts progressively deviate from geometrical figures; it is only, therefore, in the lowest vertebrated class that we find teeth in the form of perfect cubes, and of prisms or plates with three sides (as in *myletes*), four sides (as in *scarus*), five or six sides (as in *myliobates*, Fig. 60). The cone is the most common form in fishes; such teeth may be slender, sharp-pointed, and so minute, numerous, and closely aggregated, as to resemble the plush or pile of velvet. These are called "villiform teeth" (*dentes villiformes*, Lat.; *dents en velours*, Fr.). All the teeth of the perch are of this kind. When the teeth are equally fine and numerous, but longer, these are called "ciliiform" (*dentes ciliiformes*); when the teeth are similar to but rather stronger than these, they are called "setiform" (*dentes setiformes*, Lat.; *dents en brosse*, Fr.); the teeth in the upper jaw of the pike (Fig. 59) are of this kind. Conical teeth, as close-set and sharp-pointed as the villiforme teeth, but of larger size, are called "rasp-teeth" (*dentes raduliformes*, Lat.; *dents en rape*, or *en cardes*, Fr.); the pike presents such teeth on the back part of the vomer. The teeth of the sheat-fish (*Silurus glanis*) present

Fig. 60.

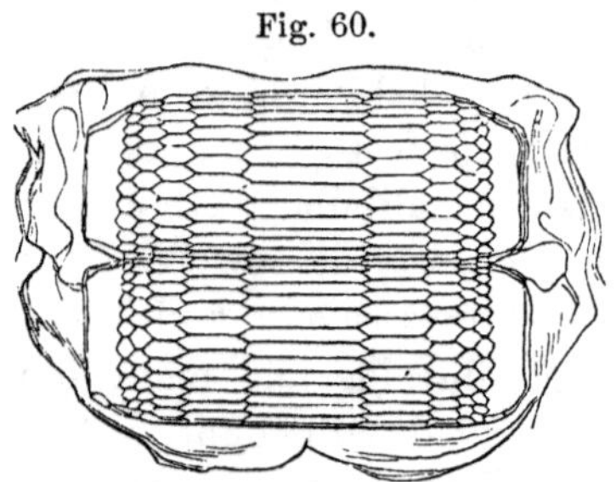

JAWS AND TEETH OF THE STING-RAY (*Myliobates*).

all the gradations between the villiform and raduliform types. Setiform teeth are common in the fishes thence called Chætodonts; in the genus *Citharinæ* they bifurcate at their free extremities; in the genus *Platax* they end there in three diverging points, and the cone here merges into the long and slender cylinder. Sometimes the cone is compressed into a slender trenchant blade: and this may be pointed and recurved, as in the *muræna;* or barbed, as in *trichiurus* and some other Scomberoides; or it may be bent upon itself, like a tenter-hook, as in the fishes thence called *Goniodonts.* In the bonito may be perceived a progressive thickening of the base of the conical teeth; and this being combined in other predatory fishes with increased size and recurved direction, they then resemble the laniary or canine teeth of carnivorous quadrupeds, as we see in the large teeth of the pike (Fig. 59), in the lophius, and in certain sharks.

The anterior diverging grappling teeth of the wolf-fish (Fig. 61), *i*, form stronger cones; and by progressive blunting, flattening, and expansion of the apex, observable in different fishes, the cone gradually changes to the thick and short cylinder, such as is seen in the back teeth of the wolf-fish, *m*, and in similar grinding and crushing teeth in other genera, whether the fishes be feeders on sea-weeds, or on crustaceous and testaceous animals. The grinding surface of these short cylindrical teeth may be convex, as in the sheep's-head fish (*Sargus*); or flattened, as in the pharyngeal teeth of the wrasse (*Labrus*). Sometimes the hemispheric teeth are so numerous, and spread over so broad a surface, as to resemble a pavement, as in the pharyngeal bones of the wrasse; or they may be so small, as well as numerous, as to give a granulated surface to the part of the mouth to which they are attached, when they are called, in ichthyology, *dentes graniformes*.

A progressive increase of the transverse over the vertical diameter may be traced in the molar teeth of different fishes, and sometimes in those of the same individual, as in *labrus*, until the cylindrical form is exchanged for that of the depressed plate. Such dental plates (*dentes lamelliformes*) may be formed not only circular, but elliptical, oval, semilunar, sigmoid, oblong, or even square, hexagonal, pentagonal, or triangular; and the grinding surface may present various and beautiful kinds of sculpturing. The broadest and thinnest lamelliform teeth are those that form the complex grinding tubercle of the diodon.

In the sharks and rays, the teeth are supported by the upper and lower jaws, as in most quadrupeds; but many other fishes have teeth growing from the roof of the mouth, from the surface of the tongue, from the bony hoops or arches supporting the gills, and some have them developed from the bone of the nose and the base of the skull. In the carp and tench the teeth are confined to this latter unusual position, and to a pair of bones, called "pharyngeal," which circumscribe the back outlet of the mouth.

Fishes exhibit, moreover, a greater range of variety in the mode of attachment of the teeth than any other class of animals. In the sharks, and the singular fish called the "angler," the teeth are movable, their base being tied by ligaments to the jaw. In the angler the ligaments are so inserted that they do not permit the teeth to be bent outwards beyond the vertical position, but yield to pressure in the contrary direction, by which the point of the tooth may be directed towards the back of the mouth; the instant, however, that the pressure is remitted, the tooth returns through the elasticity of the bent ligaments, as by the action of a spring, to its usual erect position; the deglutition of the prey of this voracious fish is thus facilitated, and its escape prevented. The broad and generally bifurcate bony base of the teeth of sharks is attached by ligaments to the semi-ossified crust of the cartilaginous jaws; but they have no power of erecting or depressing the teeth at will.

The teeth of the sphyræna are examples of the ordinary implantation in sockets, with the addition of a slight anchylosis of the base of the fully-formed tooth with the alveolar walls; and the compressed rostral teeth of the saw-fish are deeply implanted in sockets; the hind margin of their base is grooved, and a corresponding ridge from the back part of the socket fits into the groove, and gives additional fixation to the tooth.

The singular and powerfully developed dental system

of the wolf-fish (*Anarrhicas lupus*, Fig. 61) has been a subject of interest to many anatomists. Most of the teeth are powerful crushers; some present the laniary type, with the apices more or less recurved and blunted by use, and consist of strong cones, spread abroad, like grappling-hooks, at the anterior part of the mouth, *i*, *i*.

The premaxillary teeth, 22, *i*, are all conical, and arranged in two rows; there are two, three, or four in the exterior row, at the mesial half of the bone, which are the largest; and from six to eight smaller teeth are irregularly arranged behind. There are three large, strong, diverging laniaries at the anterior end of each premandibular bone, and immediately behind these an irregular number of shorter and smaller conical teeth, which gradually exchange this form for that of large obtuse tubercles, *m*, *m*; these extend backwards, in a double alternate series, along a great part of the alveolar border of the bone, and are terminated by two or three smaller teeth in a single row, the last of which again presents the conical form. Each palatine bone, 20, supports a double row of teeth, the outer ones being conical and straight, and from four to six in number; the inner ones two, three, or four in number, and tuberculate. The lower surface of the vomer, 13, is covered by a double irregularly alternate series of the same kind of large tuberculate crushing teeth as those at the middle of the premandibular bone. Thus the inside of the mouth appears to be paved with teeth, by means of which the wolf-fish can break in pieces the shells of whelks and lobsters, and effectually disengage the nutritious animal parts from them. All the teeth are anchylosed to more or less developed alveolar eminences of bones. From the enormous power of the muscles of the jaws, and the strength of the shells which are cracked and crushed by the teeth, their fracture and displacement must obviously be no unfrequent occur-

Fig. 61.

TEETH OF THE WOLF-FISH (*Anarrhicas*).

rence; and most specimens of the jaws of the wolf-fish exhibit some of the teeth either separated at this line of imperfect anchylosis, or, more rarely, detached by fracture of the supporting osseous alveolar process.

Thus, with reference to the main and fundamental tissue of tooth, we find not fewer than six leading modifications in fishes.

Hard or true dentine—Sparoids, labroids, lophius, balistes, pycnodonts, prionodon, sphyræna, megalichthys, rhizodes, diodon, scarus;

Osteodentine—Cestracion, acrodus, lepidosiren, ctenodus, hybodus, percoids, sciænoids, cottoids, gobioids, sharks, and many others;

Vasodentine—Psammodus, chimæroids, pristis, myliobates;

Plicidentine—Lophius, holoptychius, bothriolepis; and

Dendrodentine—Dendrodus;

Besides the compound teeth of the scarus and diodon.

One structural modification may prevail in some teeth, another in other teeth, of the same fish; and two or more modifications may be present in the same tooth, arising from changes in the process of calcification and a persistency of portions or processes of the primitive vascular pulp or matrix of the dentine.

The dense covering of the beak-like jaws of the parrot-fishes (*Scari*) consists of a stratum of prismatic denticles, standing almost vertically to the external surface of the jaw bone; this peculiar armature of the jaws is adapted to the habits and exigences of a tribe of fishes which browse upon the lithophytes that clothe, as with a richly tinted carpet, the bottom of the sea, just as the ruminant quadrupeds crop the herbage of the dry land.

The irritable bodies of the gelatinous polypes, which constitute the food of these fishes, retreat, when touched, into their star-shaped stony shells, and the *scari* consequently require a dental apparatus strong enough to break off or scoop out these calcareous recesses. The jaws are, therefore, prominent, short, and stout, and the exposed portions of the premaxillaries and premandibulars are incased by a complicated dental covering. The polypes and their cells are reduced to a pulp by the action of the pharyngeal jaws and teeth that close the posterior aperture of the mouth.

There is a close analogy between the dental mass of the scarus and the complicated grinders of the elephant, both in form, structure, and in the reproduction of the component denticles in horizontal succession. But in the fish, the complexity of the triturating surface is greater than in the mammal, since, from the mode in which the wedge-shaped denticles of the scarus are implanted upon, and anchylosed to, the processes of the supporting bone, this likewise enters into the formation of the grinding surface when the tooth is worn down to a certain point.

The proof of the efficacy of the complex masticatory apparatus above described, is afforded by the contents of the alimentary canal of the scari. Mr. Charles Darwin, the accomplished naturalist and geologist, who accompanied Captain Fitzroy, R. N., in the circumnavigatory voyage of the Beagle, dissected several parrot-fishes soon after they were caught, and found the intestines laden with nearly pure chalk, such being the nature of their excrements; whence he ranks these fishes among the geological agents to which is assigned the office of converting the skeletons of the lithophytes into chalk.

The most formidable dentition exhibited in the order of osseous fishes is that which characterizes the sphyræna, and some extinct fishes allied to this predatory genus. In the great barracuda of the southern shores of the United States (*Sphyræna barracuda*, Cuv.), the lower-jaw contains a single row of large, compressed, conical, sharp-pointed, and sharp-edged teeth, resembling the blades of lancets, but stronger at the base; the two anterior of these teeth are twice as long as the rest, but the posterior and serial teeth gradually increase in size towards the back part of the jaw; there are about twenty-four of these piercing and cutting teeth in each premandibular bone. They are opposed to a double row of similar teeth in the upper-jaw, and fit into the interspace of these two rows when the mouth is closed. The outermost row is situated on the intermaxillary, the innermost on the palatine bones; there are no teeth on the vomer or superior maxillary bones. The two anterior teeth in each premaxillary bone equal the opposite pair in the lower-jaw in size; the posterior teeth are serial, numerous, and of small size; the second of the two anterior large premaxillary teeth is placed on the inner side of the commencement of the row of small teeth, and is a little inclined backwards. The retaining power of all the large anterior teeth is increased by a slight posterior projection, similar to the barb of a fish-hook, but smaller. The palatine bones contain each nine or ten lancet-shaped teeth, somewhat larger than the posterior ones of the lower-jaw. All these teeth afford good examples of the mode of attachment by implantation in sockets, which has been denied to exist in fishes.

The loss or injury to which these destructive weapons are liable, in the conflict which the sphyræna wages with its living and struggling prey, is repaired by an uninterrupted succession of new pulps and teeth. The existence of these is indicated by the foramina, which are situated immediately posterior to, or on the inner margin of, the sockets of the teeth in place; these foramina lead to alveoli of reserve, in which the crowns of the new teeth in different stages of development are loosely imbedded. It is in this position of the germs of the teeth that the sphyrænoid fishes, both recent and fossil, mainly differ, as to their dental characters, from the rest of the scomberoid family, and proportionally approach the sauroid type.

In all fishes the teeth are shed and renewed, not once only, as in mammals, but frequently during the whole course of their lives. The maxillary dental plates of lepidosiren, the cylindrical dental masses of the chimæroid and edaphodont fishes, and the rostral teeth of the saw-fish (if these modified dermal spines may be so called) are, perhaps, the sole examples of "permanent teeth" to be met with in the whole class. In the great majority of fishes, the germs of the new teeth are developed like those of the old, from the free surface of the buccal membrane throughout the entire period of succession; a circumstance peculiar to the present class. The angler, the pike, and most of our common fishes, illustrate this mode of dental reproduction; it is very conspicuous in the cartilaginous fishes (Fig. 60, *e. g.*), in which the whole phalanx of their numerous teeth is ever marching slowly forwards in rotatory progress over the alveolar border of the jaw, the teeth being successively cast off as they reach the outer margin, and new teeth rising from the mucous membrane behind the rear rank of the phalanx.

This endless succession and decadence of the teeth, together with the vast number in which they often coexist in the same fish, illustrate the law of "vegetative or irrelative repetition," as it manifests itself on the first introduction of new organs in the animal kingdom, under which light we must view the above-described organized and calcified preparatory instruments of digestion in the lowest class of the vertebrate series.

DENTAL SYSTEM OF REPTILES.

In the class reptilia an entire order (*Chelonia*), including the tortoises, terrapenes, and turtles, are devoid of teeth; but the jaws in these edentulous reptiles are covered by a sheath of horn, which in some species is of considerable thickness and density; its working surface is trenchant in the carnivorous species, but is variously sculptured and adapted for both cutting and bruising in the vegetable feeders. No species of toad possesses teeth; neither have the jaws the compensatory covering above described in the chelonians. Frogs have teeth in the upper but not in the lower jaw. Newts and salamanders have teeth in both jaws, and also upon the palate; and teeth are found in the latter situation as well as on the jaws in most serpents and in the iguana lizard. In most other lizards and in crocodiles, the teeth are confined to the jaws: in the former they are cemented or anchylosed to the jaw; in the latter they are implanted in sockets.

The existing lizards exhibit many modifications in the form of the teeth, according to the nature of the food. They are pointed with sharp cutting edges in the great carnivorous monitor (*Varanus*), and are obtuse and rounded like paving-stones in the herbivorous or mixed feeding scinks, called, on account of the shape of the teeth, *cyclodus*. The gigantic extinct lizards showed similar modifications of their teeth. The megalosaurus had teeth which combined the properties of the knife, the sabre, and the saw (Fig. 62). When first protruded above the gum, the apex of the tooth presented a double cutting edge of serrated enamel; its position and line of action were nearly vertical, and its form, like that of the two-edged sword, cutting equally on each side. As the tooth advanced in growth, it became curved backwards in the form of a pruning-knife, and the edge of serrated enamel was continued downwards to the base of the concave and cutting side of the tooth; whilst on the other side a similar edge descended but a short distance from the point, and the convex part of the tooth became blunt and thick, as the back of a knife is made thick for the purpose of producing strength. In a tooth thus formed for cutting along its concave edge, each movement of the jaw combined the power of the knife and the saw. The backward curvature of the full-grown teeth enables them to retain, like barbs, the prey which they had penetrated.

Fig. 62.

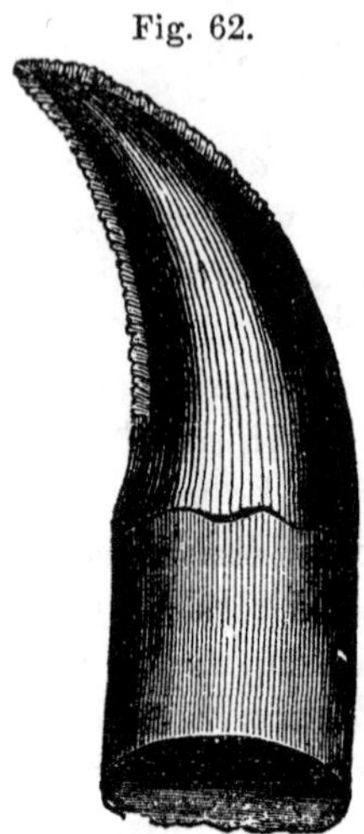

TOOTH OF THE MEGALOSAURUS.

In the iguanodon—the gigantic contemporary of the megalosaurus—the crown of the teeth (Fig. 63) was so shaped that, after the apex became worn down, it presented a broad and nearly horizontal surface, exposing dental substances of four different degrees of density, viz: a ridge of enamel along the outer border of the crown; a layer of hard or unvascular dentine

Fig. 63.

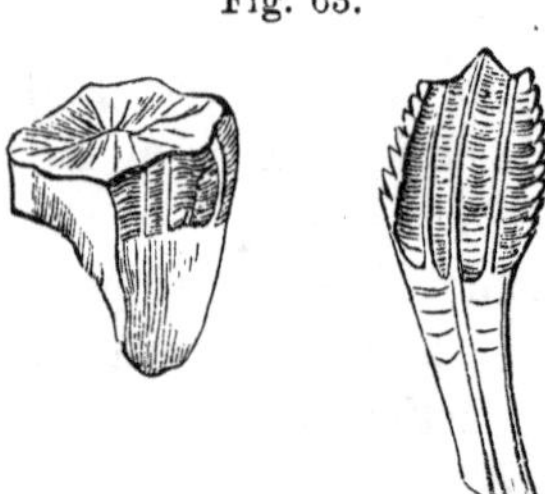

NEW-FORMED AND WORN TEETH OF THE IGUANODON.

next to this; a layer of softer vascular dentine forming the inner half of the crown; and a portion of firm osteo-dentine in the middle of the grinding surface, formed by the ossified remnant of the tooth-pulp. The series of complex teeth, so constructed, seems to have been admirably adapted to the cropping and comminution of such tough vegetable food as the *clathrariæ* and similar now extinct plants, the fossil remains of which are found buried with those of the iguanodon. No existing reptile now presents so complicated a structure of the tooth in relation to vegetable food. The still more complex, and indeed marvellous structure of the teeth of the extinct gigantic lizard-like toad, called *Labyrinthodon*, has been already noticed (Fig. 56, p. 236). But, perhaps, the most singular dental structure yet found in the ancient members of the class Reptilia, is that presented by certain species of fossil found in South Africa, and probably from a geological formation nearly as old as our coal strata. I have called them "Dicynodonts," from their dentition being reduced to one long and large canine tooth on each side of the upper jaw. As these teeth give, at first sight, a character to the jaws like that which the long poison-fangs give, when erected, to the jaws of the rattlesnake, I shall briefly notice their characters before entering upon the description of the more normal saurian dentition.

Fig. 64 gives a reduced side view of the skull and teeth of the *Dicynodon lacerticeps*.

The maxillary bone, 21, is excavated by a wide and deep alveolus, with a circular area of half an inch, and lodges a long and strong, slightly curved, and sharp-pointed canine tooth or tusk, which projects about two-thirds of its length from the open extremity of the socket.

The direction of the tusks is forwards, downwards, and very slightly inwards; the two converging in the descent

Fig. 64.

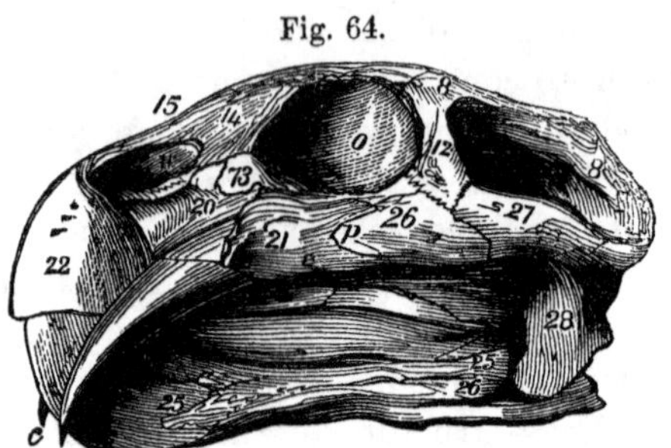

SKULL AND TUSKS OF *Dicynodon lacerticeps.*

along the outer side of the compressed symphysis of the lower jaw, *cc.* The tusk is principally composed of a body of compact unvascular dentine. The base is excavated by a wide conical pulp-cavity, *p*, with the apex extending to about one-half of the implanted part of the tusk, and a linear continuation extending along the centre of the solid part of the tusk.

Until the discovery of the rhynchosaurus, this edentulous and horn-sheathed condition of the jaws was supposed to be peculiar to the chelonian order among reptiles; and it is not one of the least interesting features of the dicynodonts of the African sandstones, that they should repeat a chelonian character hitherto peculiar amongst lacertians, to the above-cited remarkable extinct edentulous genus of the new red sandstone of Shropshire; but our interest rises almost to astonishment when, in a saurian skull, we find, superadded to the horn-clad mandibles of the tortoise, a pair of tusks, borrowed, as it were, from the mammalian class, or rather foreshadowing a structure which, in the existing creation, is peculiar to certain members of the highest organized warm-blooded animals.

In the other reptilia, recent or extinct, which most nearly approach the mammalia in the structure of their teeth, the difference characteristic of the inferior and cold-blooded class is manifested in the shape, and in the system of shedding and succession of the teeth; the base of the implanted teeth seldom becomes consolidated, never contracted to a point, as in the fangs of the simple teeth of mammalia, and at all periods of growth one or more genus of teeth are formed within or near the base of the tooth in use, prepared to succeed it, and progressing towards its displacement. The dental armature of the jaws is kept in serviceable order by uninterrupted change and succession; but the forming organ of the individual tooth is soon exhausted, and the life of the tooth itself may be said to be comparatively short.

If one of the conical, sharp-pointed, and two-edged teeth of the Gangetic crocodile, called "garrhial," by the Hindoos, be extracted, its base will be found hollow, and

Fig. 65.

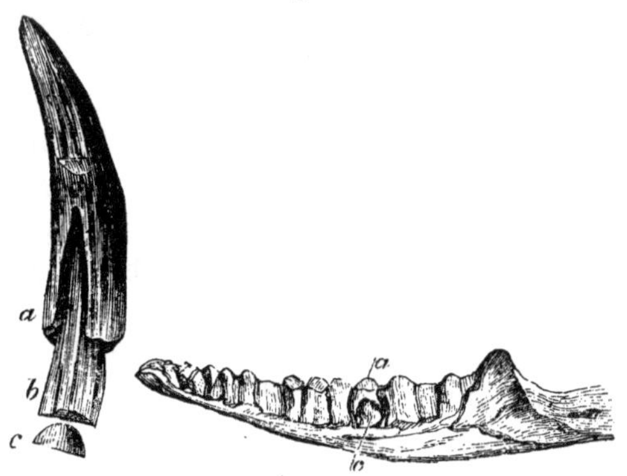

TOOTH, WITH GERMS OF SUCCESSORS, OF THE GARRHIAL (*Gavialis gangeticus*).

partly absorbed or eaten away, as at *a*, Fig. 65; and within the cavity will be seen the half-formed succeeding tooth, *b;* at the base of which may probably be found the beginning or germ, *c*, of the successor of that tooth; all the teeth in the crocodile tribe being pushed out and replaced in the vertical direction by new teeth, as long as they live. The individual teeth increase in size as the animal grows; but the number of teeth remains the same from the period when the crocodile quits the egg to the attainment of its full size and age. No sooner has the young tooth penetrated the interior of the old one, than another germ begins to be developed from the angle between the base of the young tooth and the inner alveolar process, or in the same relative position as that in which its immediate predecessor began to rise; and the processes of succession and displacement are carried on, uninterruptedly, throughout the long life of these cold-blooded carnivorous reptiles. The fossil jaws of the extinct crocodiles demonstrate that the same law regulated the succession of the teeth at the ancient epochs when they prevailed in greatest numbers, and under the most varied specific modifications, as at the present day, when they are reduced to a single family.

The most complex condition of the dental system in the reptile class is that which is presented by the poisonous serpents, in which certain teeth are associated with the tube or duct of a poison-bag and gland.

These teeth, called "poison-fangs," are confined to those bones of the upper jaw called "maxillary," and are usually single; or, when more, one only is connected with the poison-apparatus, and the others are either simple teeth, or preparing to take the place of the poison-fang.

Fig. 66.

POISON-FANG OF RATTLE-SNAKE (magnified).

To give an idea of the structure of this tooth, we may suppose a simple slender tooth, like that of a boa-constrictor, to be flattened, and its edges then bent towards each other and soldered together so as to form a tube, open at both ends, and inclosing the end of the poison-duct. Such a tooth is represented at Fig. 66, where A is the oblique opening penetrated by the duct, and *v* the narrower fissure by which the venom escapes.

The duct which conveys the poison, although it runs through the centre of the tooth, is really on the outside of the tooth. The bending of the dentine about it begins a little beyond the base of the tooth, where the poison-duct rests in a slight groove or longitudinal indentation on the convex side of the fang; as it proceeds, it sinks deeper into the substance of the tooth, and the sides of the groove meet and seem to coalesce, so that the trace of the inflected fold ceases, in some species, to be perceptible to the naked eye; and the fang appears, as it is commonly described, to be perforated by the duct of the poison-gland.

Fig. 67.

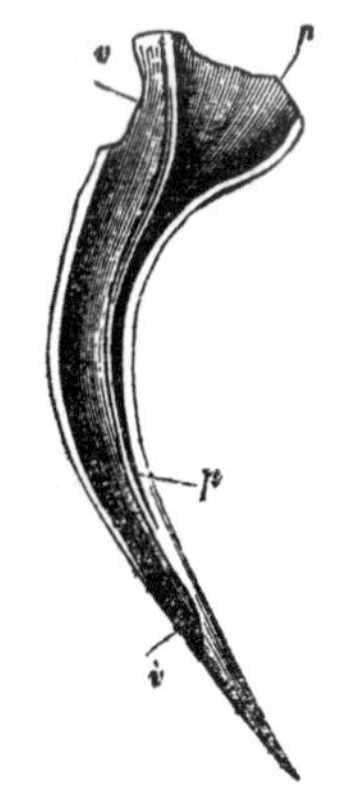

SECTION OF A POISON-FANG—RATTLESNAKE.

In the viper, the line of union may be seen as marked at *v*, Fig. 66; and when such a tooth is carefully divided lengthwise, as in Fig. 67, the true pulp-cavity in the substance of the tooth is seen, as at *p*, *p*, to terminate in a point; and the poison-canal, as at *u*, *v*, to run along the forepart of the singularly modified tooth. This tooth is soldered to the maxillary bone (Fig. 68), which rotates so as to keep the tooth laid flat in the mouth at ordinary times, and to erect it when the deadly blow is about to be struck. The head of the snake is raised, drawn back, and the fangs, erect, and exposed by the widely open mouth, are struck, by the force of the powerful muscles of the head and neck, into the surface aimed at, the poison-bags at the same moment are squeezed, and their contents driven through the canal in the tooth into the wound. And here may be noticed the advantage of having the solid point of the tooth prolonged beyond the outlet of the poison-canal, and not weakened by its continuation to the apex.

Fig. 68.

SKULL OF A RATTLESNAKE (*Crotalus horridus*).

DENTAL SYSTEM OF MAMMALS.

The class *Mammalia*, like those of *Reptilia* and *Pisces*, includes a few genera and species that are devoid of teeth; the true ant-eaters (*myrmecophaga*), the scaly ant-eaters or pangolins (*manis*), and the spiny monotrematous ant-eater (*echidna*), are examples of strictly edentulous mammals. The ornithorhynchus has horny teeth, and the whales (*balæna* and *balænoptera*) have transitory embryonic calcified teeth, succeeded by whalebone substitutes in the upper jaw.

The female narwhal seems to be edentulous, but has the germs of two tusks in the substance of the upper jaw-bones; one of these becomes developed into a large and conspicuous weapon in the male narwhal, whence the name of its genus, of *monodon*, meaning single tooth. In another cetacean, the great bottle-nose or hyperoodon, the teeth are reduced in the adult to two in number, whence the specific name *H. bidens*, but they are confined to the lower jaw.

The elephant has never more than one entire molar, or parts of two, in use on each side of the upper and lower jaws; to which are added two tusks, more or less developed, in the upper jaw.

Some rodents, as the Australian water-rats (*Hydromys*), have two grinders on each side of both jaws; which, added to the four cutting-teeth in front, make twelve in all; the common number of teeth in this order is twenty, but the hares and rabbits have twenty-eight each.

The sloth has eighteen teeth. The number of teeth, thirty-two, which characterizes man, the apes of the old world, and the true ruminants, is the average one of the class mammalia; but the typical number is forty-four.

The examples of excessive number of teeth are presented in the order Bruta, by the priodont armadillo, which has ninety-eight teeth; and in the cetaceous order by the cachalot, which has upwards of sixty teeth, though most of them are confined to the lower jaw; by the common porpoise, which has between eighty and ninety teeth; by the Gangetic dolphin, which has one hundred and twenty teeth; and by the true dolphins (*delphinus*), which have from one hundred to one hundred and ninety teeth, yielding the maximum number in the class Mammalia.

Form.—Where the teeth are in excessive number, as in the species above cited, they are small, equal, or subequal, and usually of a simple conical form.

In most other mammalia, particular teeth have special

forms for special uses: thus, the front teeth, from being commonly adapted to effect the first coarse division of the food, have been called cutters or incisors; and the back teeth, which complete its comminution, grinders or molars; large conical teeth situated behind the incisors, and adapted by being nearer the insertion of the biting muscles to act with greater force, are called holders, tearers, laniaries, or, more commonly, canine teeth, from being well developed in the dog and other carnivora.

Molar teeth, which are adapted for mastication, have either tuberculate, or transversely ridged, or flat summits, and usually are either surrounded by a ridge of enamel, or are traversed by similar ridges arranged in various patterns.

The large molars of the capybara and elephant have the crown cleft into a numerous series of compressed transverse plates, cemented together side by side.

The teeth of the mammalia have usually so much more definite and complex a form than those of fishes and reptiles, that three parts are recognized in them, viz: the "fang," the "neck," and the "crown." The fang or root (*radix*) is the inserted part; the crown (*corona*) the exposed part; and the constriction which divides these is is called the neck (*cervix*).

FIXATION.—It is peculiar to the class mammalia to have teeth implanted in sockets by two or more fangs; but this can only happen to teeth of limited growth, and generally characterizes the molars and premolars; perpetually growing teeth require the base to be kept simple and widely excavated for the persistent pulp. In no mammiferous animal does anchylosis of the tooth with the jaw constitute a normal mode of attachment. Each tooth has its particular socket, to which it firmly adheres by the close coadaptation of their opposed surfaces, and by the firm adhesion of the alveolar periosteum to the organized cement which invests the fang or fangs of the tooth.

True teeth implanted in sockets are confined, in the mammalian class, to the maxillary, premaxillary, and mandibular, or lower maxillary bones, and form a single row in each. They may project only from the premaxillary bones, as in the narwhal, or only from the lower maxillary bone, as in ziphius; or be apparent only in the lower maxillary bone, as in the cachalot; or be limited to the superior and inferior maxillaries, and not present in the premaxillaries, as in the true *peccora* (cow, sheep), and most *bruta* (sloth, armadillo) of Linnæus. In general, teeth are situated in all the bones above mentioned. In man, where the premaxillaries early coalesce with the maxillary bones, where the jaws are very short, and the crowns of the teeth are of equal length, there is no vacant space in the dental series of either jaw, and the teeth derive some additional fixity by their close apposition and mutual pressure. No inferior mammal now presents this character; but its importance, as associated with the peculiar attributes of the human organization, has been somewhat diminished by the discovery of a like contiguous arrangement of the teeth in the jaws of a few extinct quadrupeds, *e. g.* anoplotherium, nesodon, and dichodon.

STRUCTURE.—The teeth of the mammalia usually consist of hard unvascular dentine, defended at the crown by an investment of enamel, and everywhere surrounded by a coat of cement. The coronal cement is of extreme tenuity in man, quadrumana, and the terrestrial carnivora; it is thicker in the herbivora, especially in the complex grinders of the elephant. Vertical folds of enamel and cement penetrate the crown of the tooth in the ruminants, and in most rodents and pachyderms, characterizing by their varions forms the genera of the last two orders.

The teeth of the sloths, armadillos, and sperm-whales have no true enamel. The tusks of the narwhal, walrus, and elephant consist of modified dentine, which in the last great proboscidian animal is properly called "ivory," and is covered by cement.

The forming-organ of a mammalian tooth consists, as in the lower classes, of a pulp and a capsule. The substance of the pulp is converted into the "dentine;" that of the capsule into the "cement." Where enamel is to be added, a peculiar organ is formed on the inner surface of the capsule, which arranges the hardening material into the form, and of the density, characteristic of enamel. This substance is so hard in the tooth of the hippopotamus, as to "strike fire" like flint with steel. The whole forming-organ is called "matrix."

The matrix of certain teeth does not give rise during any period of their formation to the germ of a second tooth, destined to succeed the first; this tooth, therefore, when completed and worn down, is not replaced. The sperm whales, dolphins, and porpoises are limited to this simple provision of teeth. In the armadillos and sloths, the want of germinative power, as it may be called, in the matrix is compensated by the persistence of the matrix, and by the uninterrupted growth of the teeth.

In most other mammalia, the matrix of the first developed tooth gives origin to the germ of a second tooth, which sometimes displaces the first, sometimes takes its place by the side of the tooth from which it has originated. All those teeth which are displaced by their progeny are called temporary, deciduous, or milk teeth; the mode and direction in which they are displaced and succeeded, viz:

from above downwards in the upper, from below upwards in the lower jaw: in both jaws vertically—are the same as in the crocodile; but *the process is never repeated more than once in any mammiferous animal.* A considerable proportion of the dental series is thus changed; the second or permanent teeth having a size and form as suitable to the jaws of the adult as the displaced temporary teeth were adapted to those of the young animal.

The permanent teeth, which assume places not previously occupied by deciduous ones, are always the most posterior in their position, and generally the most complex in their form.

The term "molar," or "true molar," is restricted to these teeth; the teeth between them and the canines are called "premolars;" they push out the milk-teeth that precede them, and are usually of smaller size and simpler form than the true molars. They are called "bicuspids" in human anatomy.

Fig. 69.

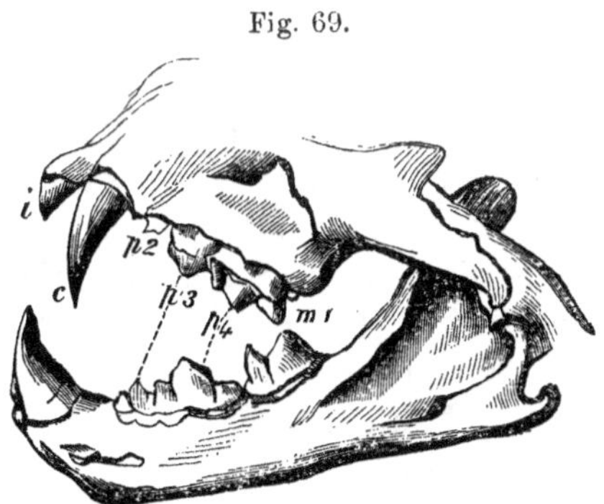

JAWS AND TEETH OF THE LION.

Thus the class mammalia, in regard to the times of formation and the succession of their teeth, have been divided into two groups: the *monophyodonts*,[1] or those that generate a single set of teeth; and the *diphyodonts*,[2] or those that generate two sets of teeth.

I proceed next to notice the principal modifications of the teeth, as they are adapted to carnivorous, herbivorous, or mixed feeding habits in the diphyodont mammalia.

The lion may be taken as the type of the flesh-feeders (Fig. 69).

The largest and most conspicuous teeth in this and other feline quadrupeds are the "canines," *c*; they are of great strength, deeply implanted in the jaw, with the fang thicker and longer than the enamelled crown: this part is conical, slightly recurved, sharp-pointed, convex in front, almost flat on the inner side, and with a sharp edge behind. The lower canines pass in front of the upper ones when the mouth is closed.

The incisors, six in number on both jaws, form a transverse row; the outermost above, *i*, is the largest, resembling a small canine: the intermediate ones have broad and thick crowns indented by a transverse cleft.

The first upper premolar, *p* 2, is rudimental: there is no answerable tooth in the lower jaw. The second, *p* 3, in both jaws has a strong conical crown supported on two fangs. The third upper tooth, *p* 4, has a cutting or trenchant crown, divided into three lobes, the last being the largest, and with a flat inner side, against which the cutting tooth, *m* 1, in the lower jaw works, like a scissor-blade. Behind, and on the inner side of the upper tooth, *p* 4, there is a small tubercular tooth, *m* 1. A glance at the long and strong, sub-compressed, trenchant, and sharp-pointed canines, suffices to appreciate their pecu-

[1] *μόνος*, once; *φύω*, I generate; *οδους*, tooth.

[2] *δις*, twice; *φύω* and *οδους*.

liar adaptation to seize, to hold, to pierce and lacerate a struggling prey. The jaws are strong, but shorter than in other carnivora, and with a concomitant reduction in the number of the teeth: thus, the canines are brought nearer to the insertion of the very powerful biting muscles (called "temporal" and "masseter"), which work them with proportionally greater force. The use of the small pincer-shaped incisor teeth is to gnaw the soft gristly ends of the bones, and to tear and scrape off the tendinous attachments of the muscles and the periosteum. The compressed trenchant blades of the sectorial teeth play vertically upon each other's sides, likes the blades of scissors, serving to cut and coarsely divide the flesh; and the form of the joint of the lower jaw almost restricts its movement to the vertical direction, up and down. The wide and deep zygomatic arches, and the high crests of bone upon the skull, concur in completing the carnivorous physiognomy of this most formidable of the feline tribe.

The dentition of the hyena assumes those characteristics which adapt it for the peculiar food and habits of the adult. The main modification is the great size and strength of the molars as compared with the canines, and more especially the thick and strong conical crowns of the second and third premolars in both jaws, the base of the cone being belted by a strong ridge which defends the subjacent gum. This form of tooth is especially adapted for gnawing and breaking bones, and the whole cranium has its shape modified which work the jaws and teeth in this operation.

The strength of the hyena's jaw is such that, in attacking a dog, he begins by biting off his leg at a single snap. Adapted, however, to obtain its food from the coarser parts of animals which are left by the nobler beasts of prey, the hyena chiefly seeks the dead carcass, and bears the same relation to the lion which the vulture does to the eagle. The hyena cracks, crushes, and devours the bones as well as the softer parts of the animals it preys upon. In consequence of the quantity of bones which enter into its food, the excrements consist of solid balls of a yellowish-white color, and of a compact earthy fracture. Such specimens of the substance, known in the old "Materia Medica" by the name of "album græcum," were discovered by Dr. Buckland in the celebrated ossiferous cavern at Kirkdale. They were recognized at first sight by the keeper of a menagerie, to whom they were shown, as resembling both in form and appearance the feces of the spotted hyena; and, being analyzed by Dr. Wollaston, were found to be composed of the ingredients that might be expected in fecal matter derived from bones—viz., phosphate of lime, carbonate of lime, and a very small proportion of the triple phosphate of ammonia and magnesia. This discovery of the coprolites of the hyena formed, perhaps, the strongest of the links in that chain of evidence by which Dr. Buckland proved that the cave at Kirkdale, in Yorkshire, had been, during a long succession of years, inhabited as a den by hyenas, and that they dragged into its recesses the other animal bodies, whose remains, splintered, and bearing marks of teeth of the hyena, were found mixed indiscriminately with their own.

Fig. 70.

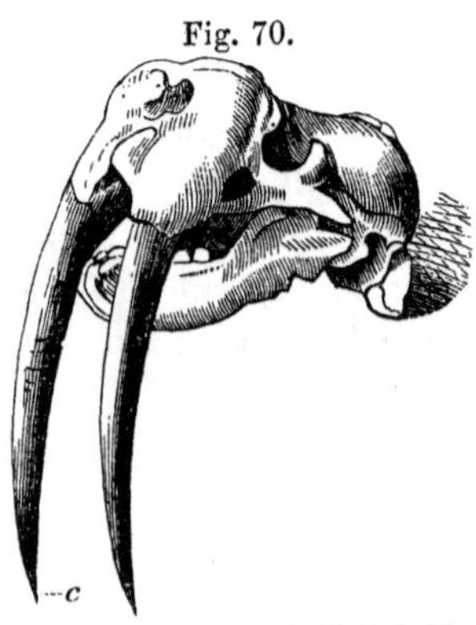

SKULL AND TEETH OF THE MORSE.

Before quitting the carnivorous order, the peculiar development of the upper canines of the morse or walrus deserve to be noticed. The staple food of this large modified seal is shell-fish, crustaceans, and sea-weed, which are pounded to a pulp by its small obtuse molar teeth. The canines (Fig. 70), *c*, exist only in the upper jaw, where they are imbedded in deep and large prominent sockets, whence they sweep down, slightly incurved, forming large and long tusks, which serve as weapons of attack and defence, and as instruments in aid of climbing the floes and hummocks of ice, amongst which the walrus passes its existence.

In the order of mammalia, called gnawers or rodents, some of which, *e. g.* the rat, are mixed feeders, but most of them herbivorous, the canine teeth are wanting in both jaws, and the incisors, reduced to two in number, are the seat of that excessive and uninterrupted growth, which makes them allied to tusks.

These incisors (Fig. 71), *i*, are curved, the upper pair

Fig. 71.

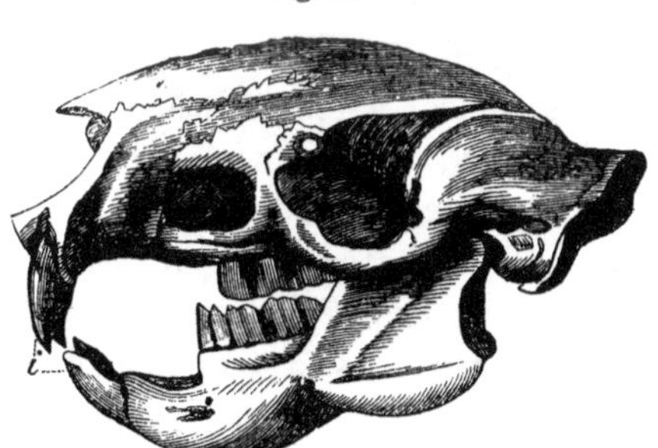

SKULL AND TEETH OF A PORCUPINE.

describing a larger part of a smaller circle, the lower ones a smaller part of a larger circle, the latter being the longest, and usually having their sockets extending from the fore to the back part of the under jaw. The tooth consists of a body of compact dentine, with a plate of enamel laid upon its anterior or convex surface, and the enamel commonly consists of two layers, of which the anterior and external one is the densest. Thus the substances of the incisor diminish in hardness from the front to the back part of the tooth. The wear and tear from the reciprocal action of the upper and lower incisors produce, accordingly, an oblique surface, sloping from a sharp anterior margin formed by the denser enamel, like that which, in a chisel, slopes from the sharp edge formed by the plate of hard steel laid on the back of that tool, whence these teeth have been called "chisel-teeth" (*dentes scalprarii*). Their growth never ceases while the animal lives, and the implanted part retains the form and size of the exposed part, and ends behind in a widely open or hollow base, which contains a long, conical, persistent forming pulp. This law of unlimited growth is unconditional, and constant exercise and abrasion are required to maintain the normal form and serviceable proportions of the scalpriform teeth of the rodents. When, by accident, an opposing incisor is lost, or when, by the distorted union of a broken jaw, the lower incisors no longer meet the upper ones, as sometimes happens to a wounded hare or rabbit, the incisors continue to grow until they project, like the tusks of the elephant, and the extremities, in the poor animal's attempts to acquire food, also become pointed like tusks. Following the curve prescribed to their growth by the form of their socket, their points often return against some part of the head, are passed through the skin, cause absorption of the bone, and perhaps again enter the mouth, rendering mastication impracticable, and causing death by starvation. In the Museum of the College of Surgeons there is a lower jaw of a beaver, in which the scalpriform incisor has, by unchecked growth, described a complete circle; the point has pierced the masseter muscle, entered the back of the mouth, and terminated close to the bottom of the socket containing its own hollow root.

The difference in the diet of the rodent quadrupeds has been alluded to; there is a corresponding difference in the mode of implantation of their molar teeth. Those which subsist on mixed food, and which, like the rats, betray a tendency to carnivorous habits, or which subsist, like squirrels, on the softer and more nutritious vegetable substances, as the kernels of nuts, suffer less rapid abrasion of the grinding teeth; a less depth of crown is, therefore, needed to perform the office of mastication during the brief period of life allotted to these active little mammals; and, as the economy of nature is manifested in the smallest particulars as well as in her grandest operations, no more dental substance is developed after the crown is formed than is requisite for the firm fixation of the tooth in the jaw.

The rodents that exclusively subsist on vegetable substances, especially of the coarser and less nutritious kinds, as herbage, foliage, and the bark and wood of trees, wear away more rapidly the grinding surface of the molar teeth; the crowns are, therefore, larger, and their growth continues by a reproduction of the formative matrix at their base in proportion as its calcified constituents, forming the working part of the tooth, are worn away. So long as this reproductive force is active, the molar tooth is implanted, like the incisor, by a long, undivided continuation of the crown. These rootless and perpetually growing molars are always more or less curved, for they derive from this form the same advantage as the incisors, in the relief of the delicate tissues of the active vascular matrix from the effects of the pressure which would otherwise have been transmitted more directly from the grinding surface; the capybara, and the Patagonian hare (*Dolichotis*), afford good examples of this more complex condition of the grinding teeth.

The variety in the pattern of the folds of enamel that penetrate the substance of the tooth, and add to its triturating power, is almost endless; but the folds have always a tendency to a transverse direction across the crown of the tooth in the rodents. This direction relates to the shape of the joint of the lower jaw, which almost restricts it to horizontal movements to and fro, during the act of mastication. In the true hoofed herbivorous animals, in which the joint of the lower jaw allows a free rotatory movement, the folds of enamel take other forms and directions, with modifications, constant in each genus, and characteristic of such.

The horse is here selected as an example of such herbivorous dentition (Fig. 72). The grinding teeth are six in number, on each side of both upper and lower jaws, with thick square crowns of great length, and deeply implanted in the sockets, those of the upper jaw being slightly curved. When the summits or exposed ends of these teeth begin to be worn down by mastication, the interblended enamel, dentine, and cement show the pattern figured in Cut 72; it is penetrated from within by a valley, entering obliquely from behind forwards, and dividing into or crossed by the two crescentic valleys, which soon become insulated. There is a large lobe at the end of the valley. The outer surface of the crown is impressed by two deep longitudinal channels. In the lower jaw the teeth are narrower transversely than in the upper jaw, and are divided externally into two convex lobes, by a median longitudinal fissure; internally they present three principal unequal convex ridges, and an anterior and posterior narrower ridge. All the valleys, fissures,

Fig. 72.

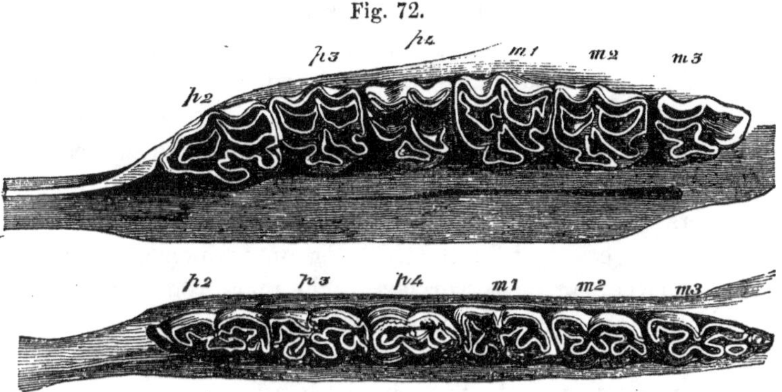

GRINDING SURFACES OF THE UPPER AND LOWER MOLARS OF A HORSE.

and folds in both upper and lower grinders are lined by enamel, which also coats the whole exterior surface of the crown. Of the series of six teeth in each jaw, the first three, *p* 2, 3, 4, are premolars, the rest, *m* 1, 2, 3, are true molars.

The canines are small in the horse, and are rudimental in the mare; the unworn crown is remarkable for the folding in of the anterior and posterior margins of enamel. The upper canine is situated in the middle of the long interspace between the incisors and molars; the lower canine is close to the outer incisor, but is distinguished by its more pointed form. The incisors are six in number in both jaws; they are arranged close together in a curve, at the end of the jaw; the crown is broad, and the contour of the biting surface, before it is much worn, approaches an ellipse. The incisors of the horse are distinguished from those of ruminants by their greater length and curvature, and from those of all other animals by the fold of enamel (Fig. 55), *a*, which penetrates the crown from its flat summit, like the inverted finger of a glove. When the tooth begins to be worn, the fold becomes an island of enamel, inclosing a cavity partly filled by cement, and partly by the substances of the food, and is called the "mark." In aged horses, the incisors are worn down below the extent of the fold, and the "mark" disappears. This cavity is usually obliterated in the first or mid incisors at the sixth year, in the second incisors at the seventh year, and in the third or outer incisors at the eighth year, in the lower jaw. The mark remains somewhat longer in the incisors of the upper jaw.

The following is the average course of development and succession of the teeth in the horse (*Equus caballus*): The summits of the first functional deciduous molar, *d*, 2, "first grinder" of veterinary authors, are usually apparent at birth; the succeeding grinder, *d* 3, sometimes arises a day or two later, sometimes together with the first. Their appearance is speedily followed by that of the first deciduous incisor—"centre nipper" of veterinarians—which usually cuts the gum between the third and sixth days. The second deciduous incisor appears between the twentieth and fortieth days, and about this time the rudimental grinder, *p* 1, comes into place, and the last deciduous molar, *d* 4, begins to cut the gum; about the sixth month the inferior lateral, or third incisors, with the deciduous canine, make their appearance. The minute canine is shed about the time that the contiguous incisor is in place, and is not retained beyond the first year. The upper deciduous canine is shed in the course of the second year. The first true molar, *m* 1, appears between the eleventh and thirteenth months. The second molar follows before the twentieth month. The first functional premolar, *p* 2, displaces the deciduous molar, *d* 2, at from two years to two years and a half old. The first permanent incisor protrudes from the gum at between two years and a half and three years. At the same period, the penultimate premolar, *p* 3, pushes out the penultimate milk molar, *d* 3, and the penultimate true molar comes into place. The last premolar displaces the last deciduous molar at between three years and a half and four years; the appearance above the gum of the last true molar, *m* 3, is usually somewhat earlier. The second incisor pushes out its deciduous predecessor about the same period. The permanent canine, or "tusk," next follows; its appearance indicates the age of four years, but it sometimes comes earlier. The third, or outer incisor, pushes out the deciduous incisor about the fifth year, but is seldom in full place before the horse is five and a half years old. Upon the rising of the third permanent incisor, or "corner nipper" of the veterinarians, the "colt" becomes a "horse," and the "filly," a "mare," in the language of the horse-dealers. After the disappearance of the "mark" in the incisors, at the eighth or ninth year the horse becomes "aged."

The most complex condition of teeth adapted to a vegetable diet is that presented by the elephant. The dentition of the genus *Elephas* includes two long tusks (Fig. 73), one in each of the intermaxillary bones, and large and complex molars (*ib.*), *m* 3, 4, and 5, in both jaws; of the latter there is never more than one wholly, or two partially, in place and use on each side at any given time,

Fig. 73.

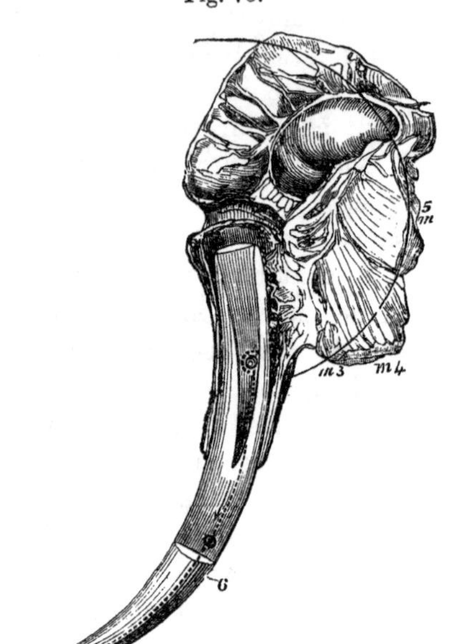

SECTION OF SKULL AND TEETH OF ELEPHANT.

for the series is continually in progress of formation and destruction, of shedding and replacement; and all the grinders succeed one another, like true molars, horizontally, from behind forward.

The total number of teeth developed in the elephant appears to be $i\frac{2-2}{0-0}, m\frac{7-7}{7-7}=32$, the two large permanent incisors being preceded by two small deciduous ones, and the number of molar teeth which follow one another on each side of both jaws being seven, or at least six, of which the last three may, by analogy, be regarded as answering to the true molars of other pachyderms.

The incisors not only surpass other teeth in size, as belonging to a quadruped so enormous, but they are the largest of all teeth in proportion to the size of the body; representing, in a natural state, those monstrous tusks of the rodents, which are the result of accidental suppression of the wearing force of the opposite teeth.

The tusks of the elephant consist chiefly of that modification of dentine that is called "ivory," and which shows, on transverse fractures or sections, striæ proceeding in the arc of a circle from the centre to the circumference, in opposite directions, and forming by their decussations curvilinear lozenges. This character is peculiar to the tusks of the proboscydian pachyderms.

In the Indian elephant, the tusks are always short and straight in the female, and less deeply implanted than in the male; she thus retaining, as usual, more of the characters of the immature state. In the male, they have been known to acquire a length of nine feet, with a basal diameter of eight inches, and to weigh one hundred and fifty pounds; but these dimensions are rare in the Asiatic species.

A mammoth's tusk has been dredged up off Dungeness which measured eleven feet in length.[1] In several of the instances of mammoth's tusks from British strata, the ivory has been so little altered as to be fit for the purposes of manufacture; and the tusks of the mammoth, which are still better preserved in the frozen drifts of Siberia, have long been collected in great numbers as articles of commerce. In the account of the mammoth's

[1] Owen's "History of British Fossil Mammalia," 8vo. 1844, p. 244.

bones and teeth of Siberia, published in the *Philosophical Transactions* for 1737, No. 446, tusks are cited which weighed two hundred pounds each, and "are used as ivory to make combs, boxes, and such other things, being but little more brittle, and easily turning yellow by weather and heat." From that time to the present there has been no intermission in the supply of ivory, furnished by the tusks of the extinct elephants of a former world.

The musket-balls and other foreign bodies which are occasionally found in ivory, are immediately surrounded by osteo-dentine in greater or less quantity. It has often been a matter of wonder how such bodies should become completely imbedded in the substance of the tusk, sometimes without any visible aperture, or how leaden bullets may have become lodged in the solid centre of a very large tusk without having been flattened. The explanation is as follows: a musket-ball, aimed at the head of an elephant, may penetrate the thin bony socket and the thinner ivory parietes of the wide conical pulp-cavity occupying the inserted base of the tusk; if the projectile force be there spent, the ball will gravitate to the opposite and lower side of the pulp-cavity, as indicated in Fig. 73. The presence of the foreign body exciting inflammation of the pulp, an irregular course of calcification ensues, which results in the deposition around the ball of a certain thickness of osteo-dentine. The pulp then resuming its healthy state and functions, coats the surface of the osteo-dentine inclosing the ball, together with the rest of the conical cavity into which that mass projects, with layers of normal ivory.

The portions of the cement-forming capsule surrounding the base of the tusk, and the part of the pulp, which were perforated by the ball in its passage, are soon replaced by the active reparative power of these highly vascular bodies. The hole formed by the ball in the base of the tusk is then more or less completely filled up by a thick coat of cement from without, and of osteo-dentine from within.

By the continued progress of growth, the ball so inclosed is carried forwards, in the course indicated by the arrow in Fig. 73, to the middle of the solidified exserted part of the tusk. Should the ball have penetrated the base of the tusk of a young elephant, it may be carried forwards by the uninterrupted growth and wear of the tusk, until that base has become the apex, and be finally exposed and discharged by the continual abrasion to which the apex of the tusk is subjected.

I had the tusk and pulp of the great elephant at the Zoological Gardens longitudinally divided, soon after the death of that animal in the summer of 1847. Although the pulp could be easily detached from the inner surface of the pulp-cavity, it was not without a certain resistance; and when the edges of the coadapted pulp and tooth were examined by a strong lens, the filamentary processes from the outer surface of the pulp could be seen stretching as they were withdrawn from the dentinal tubes before they broke. They are so minute that, to the naked eye, the detached surface of the pulp seems to be entire, and Cuvier was thus deceived in concluding that there was no organic connection between the pulp and the ivory.

The molar teeth of the elephant are remarkable for their great size, even in relation to the bulk of the animal, and for the extreme complexity of their structure. The crown, of which a great proportion is buried in the socket, and very little more than the grinding surface appears above the gum, is deeply divided into a number

24

of transverse perpendicular plates, consisting each of a body of dentine, coated by a layer of enamel, *e*, and this again by the less dense bone-like substance, *c*, which fills the interspaces of the enamelled plates, and here more especially merits the name of "cement," since it binds together the several divisions of the crown before they are fully formed and united by the confluence of their bases into a common body of dentine. As the growth of each plate begins at the summit, they remain detached, and like so many separate teeth or denticules, until their base is completed, when it becomes blended with the bases of contiguous plates to form the common body of the crown of the complex tooth, from which the roots are next developed.

Fig. 74.

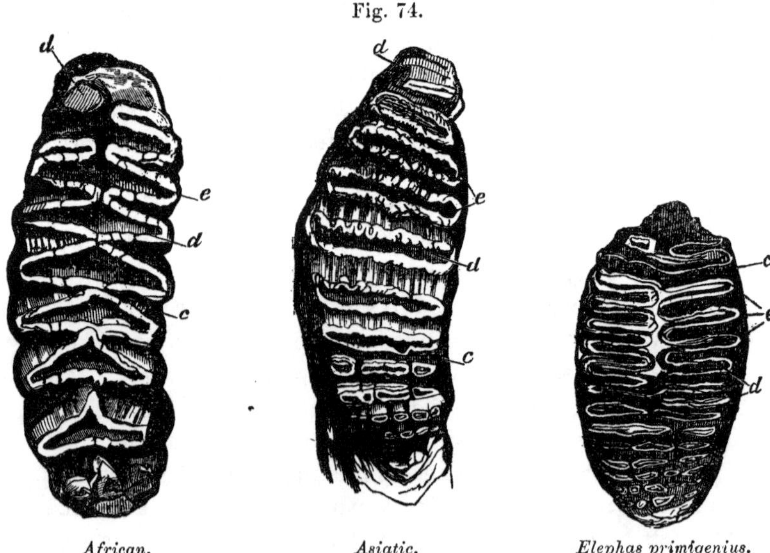

African. *Asiatic.* *Elephas primigenius.*

MOLAR TEETH OF ELEPHANTS AND THE SIBERIAN MAMMOTH.

The plates of the molar teeth of the Siberian mammoth (*Elephas primigenius*), (Fig. 74), are thinner in proportion to their breadth, and are generally a little expanded at the middle; and they are more numerous in proportion to the size of the crown than in the existing species of Asiatic elephant (*ib.*). In the African elephant (*ib.*), on the other hand, the lamellar divisions of the crown are fewer and thicker, and they expand more uniformly from the margins to the centre, yielding a lozenge-form when cut or worn transversely, as in mastication.

The formation of each grinder begins with the summits of the anterior plate, and the rest are completed in succession; the tooth is gradually advanced in position as its growth proceeds; and in the existing Indian elephant the anterior plates are brought into use before the posterior ones are formed. When the complex molar cuts the gum, the cement is first rubbed off the digital summits; then their enamel cap is worn away, and the central dentine comes into play with a prominent enamel ring; the digital processes are next ground down to their common uniting base, and a transverse tract of dentine, with its wavy border of enamel, is exposed; finally, the transverse plates themselves are abraded to their common base of dentine, and a smooth and polished tract of that substance is produced. From this basis the roots of the molar are developed, and increase in length to keep the worn crown on the grinding level, until the reproductive force is exhausted. When the whole extent of a grinder has successively come into play, its last part is reduced to a long fang, supporting a smooth and polished field of dentine, with, perhaps, a few remnants of the bottom of the enamel folds at its hinder part. When the complex molar has been thus worn down to a uniform surface, it becomes useless as an instrument for grinding the coarse vegetable substances on which the elephant subsists; it is attacked by the absorbent action, and the wasted portion of the molar is finally shed.

The grinding-teeth of the elephant progressively increase in size, and in the number of lamellar divisions, from the first to the last; they succeed each other from behind forwards, moving not in a right line, but in the arc of a circle, shown by the curved line in Fig. 73. The position of the growing tooth in the closed alveolus, *m*, 5, is almost at right angles with that in use, the grinding surface being at first directed backwards in the upper jaw, and forwards in the lower jaw, and brought by the revolving course into a horizontal line in both jaws, so that they oppose each other when developed for use. The imaginary pivot on which the grinders revolve is next their root in the upper jaw, and is next the grinding surface in the lower jaw; in both, towards the frontal surface of the skull. Viewing both upper and lower molars as one complex whole, subject to the same revolving movement, the section dividing such whole into upper and lower portions runs parallel to the curve described by the movement, the upper being the central portion, or that nearest the pivot; the lower, the peripheral portion. The grinding surface of the upper molars is consequently convex from behind forwards, and that of the lower molars concave; the upper molars are always broader than the lower ones.

The bony plate forming the sockets of the growing teeth is more than usually distinct from the body of the maxillary, and participates in this revolving course, advancing forwards with the teeth.

SUCCESSION.—As the rate of increase, both of size and in the number of the component plates of the grinding tooth, is nearly identical in both jaws, it will suffice to briefly describe the teeth and the periods at which they successively appear in the lower jaw of the Asiatic elephant.

The *first molar*, which cuts the gum in the course of the second week after birth, has a sub-compressed crown, nine lines in antero-posterior diameter, divided by three transverse clefts into four plates, the third being the broadest, and the tooth here measuring six lines across; the base slightly contracts, and forms a neck as long as the enamelled crown, but of less breadth, and this divides into an anterior and posterior, long, sub-cylindrical, diverging, but mutually incurved fangs; the total length of this tooth is one inch and a half. The corresponding upper molar cuts the gum a little earlier than the lower one: the neck of this tooth is shorter, and the two fangs are shorter, larger, and more compressed than those of the lower first molar. The first molar of the elephant is the homologue of the probably deciduous molar (Fig. 75), *d* 2, in other ungulates; it is not a mere miniature of the great molars of the mature animal, but retains, agreeably with the period of life at which it is developed, a character much more nearly approaching that of the ordinary pachydermal molar, manifesting the adherence to the more general type by the minor complexity of the crown, and by the form and relative size of the fangs. In the transverse divisions of the crown we perceive the affinity to the tapiroid type, the different links connecting which with the typical elephants are supplied by the extinct lophiodons, dinotheriums, and mastodons. The subdivision of the summits of the primary plates recalls the character of the molars, especially the smaller ones, of the phacochere in the hog tribe. As the elephant advances in age, the molars rapidly acquire their more special and complex character.

The first molars are completely in place and in full use at three months, and are shed when the elephant is about two years old.

The sudden increase and rapid development of the *second molar* may account for the non-existence of any vertical successor, or "premolar," to the former tooth, in the elephant. The eight or nine plates of the crown are formed in the closed alveolus, behind the first molar by the time this cuts the gum, and they are united with the body of the tooth, and most of them in use, when the first molar is shed. The average length of the second molar is two inches and a half, ranging from two inches to two inches and nine lines. The greatest breadth, which is behind the middle of the tooth, is from one inch to one inch three lines. There are two roots; the cavity of the small anterior one expands in the crown, and is continued into that of the three anterior plates. The thicker root supports the rest of the tooth. The second molar is worn out and shed before the beginning of the sixth year.

The *third molar* has the crown divided into from eleven to thirteen plates; it averages four inches in length, and two inches in breadth, and has a small anterior, and a very large posterior root; it begins to appear above the gum about the end of the second year, is in its most complete state and extensive use during the fifth year, and is worn out and shed in the ninth year. The last remnant of the third molar is shown at *m* 3 (Fig. 73). It is probable that the three preceding teeth are analogous to the deciduous molars, *d* 2, *d* 3, and *d* 4, in the hog (Fig. 75).

The *fourth molar* presents a marked superiority of size over the third, and a somewhat different form; the anterior angle is more obliquely abraded, giving a pentagonal figure to the tooth in the upper jaw (Fig. 73), *m* 4. The number of plates in the crown of this tooth is fifteen or sixteen, its length between seven and eight inches, its breadth three inches. It has an anterior simple and slender root supporting the first three plates, a second of larger size and bifid, supporting the four next plates, and a large contracting base for the remainder. The forepart of the grinding surface of this tooth begins to protrude through the gum at the sixth year; the tooth is worn away, and its last remnant shed, about the twentieth or twenty-fifth year. It may be regarded as the homologue of the first true molar of ordinary pachyderms (Fig. 75), *m* 1.

The *fifth molar*, with a crown of from seventeen to twenty plates, measures between nine and ten inches in length, and about three inches and a half in breadth. The second root is more distinctly separated from the first simple root than from the large mass behind. It begins to appear above the gum about the twentieth year; its duration has not been ascertained by observation, but it probably is not shed before the sixtieth year.

The *sixth molar* is the last, and has from twenty-two to twenty-seven plates; its length, or antero-posterior extent, following the curvature, is from twelve to fifteen inches; the breadth of the grinding surface rarely exceeds three inches and a half. One may reasonably conjecture that the sixth molar of the Indian elephant, if it make its appearance about the fiftieth year, would, from its superior depth and length, continue to do the work of mastication until the pondrous pachyderm had passed the century of its existence.

DEVELOPMENT.—The long-mistaken phenomena of the formation of the dental substances will be here described as they have been observed in the large teeth of the elephant; if the description be comprehended in regard to these, the most complex members of the dental system, the true theory of dental development will be readily understood in regard to all the various forms and gradations of teeth. The matrix, or formative organ of the tusk, consists of a large conical pulp, which is renewed quicker than it is converted, and thus is not only preserved, but grows, up to a certain period of the animal's life; it is lodged in the cavity at the base of the tusk; this base is surrounded by the remains of the capsule, a soft, vascular membrane of moderate thickness, which is confluent with the border of the base of the pulp, where it receives its principal vessels.

Each molar of the elephant is formed in the interior of a membranous sac—the capsule, the form of which partakes of that of the future tooth, being cubical in the first molar, oblong in the last, and rhomboidal in most of the intermediate teeth; but always decreasing in vertical extent towards its posterior end, and closed at all points, save where it is penetrated by vessels and nerves. It is lodged in an osseous cavity of the same form as itself, and usually in part suspended freely in the maxillary bone, the bony case being destined to form part of the socket of the tooth. The exterior of the membranous capsule is simple and vascular, as shown at *m* 5, Fig. 73; its internal surface gives attachment to numerous folds or processes, as in most other ungulate animals.

The dentinal pulp rises from the bottom of the capsule, or that part which lines the deepest part of the alveolus, in the form of transverse parallel plates extending towards that part of the capsule ready to escape from the socket. These plates adhere only to the bottom of the capsule; their opposite extremity is free from all adhesion. This summit is thinner than the base; it might be termed the edge of the plate; but it is notched, or divided into many digital processes. The tissue of these digitated plates is identical with that of the dentinal pulp of simple Mammalian teeth; it becomes also highly vascular at the parts where the formation of the dentine is in active progress.

Processes of the capsule descend from its summit into the interspaces of the dentinal pulp-plates, and consequently resemble them in form; but they adhere not only by their base to the surface of the capsule next the mouth, but also by their lateral margins to the sides of the capsule, and thus resemble partition-walls, confining each plate of the dentinal pulp to its proper chamber; the margin of the partition opposite its attached base is free in the interspace of the origins of the dentinal pulp-plates.

The enamel organ, which Cuvier appears to have recognized under the name of the internal layer of the capsule, is distinguishable by its light blue sub-transparent color and usual microscopic texture, adhering to the free surface of the partitions formed by the true inner layer of the capsule. Although the enamel-pulp be in close contact with the dentinal pulp prior to the commencement of the formation of the tooth, one may readily conceive a vacuity between them, which is continued uninterruptedly, in many foldings, between all the gelatinous plates of the dentinal pulp, and the partitions formed by the combined enamel-pulp and the folds of the capsule. According to the excretion view, this delicate apparatus must have been immediately subjected to the violence of being compressed in the unyielding bony box, by the deposition of the dense matters of the tooth in the hypothetical vacuity between the enamel and dentinal pulps; a process of absorption must have been conceived to be set on foot immediately that the altered condition of the gelatinous secreting organs took place; and, according to Cuvier's hypothesis, the secreting function must be supposed to have proceeded, without any irregularity or interruption, while the process of absorption was superinduced in the same part to relieve it from the effects of pressure produced by its own secretion.

The formation of the dentine commences immediately beneath the *membrana propria* of the pulp; a part which Cuvier distinctly recognized, and which he accurately traced as preserving its relative situation between the dentine and enamel throughout the whole formation of the dentine, and discernible in the completed tooth "as a very fine grayish line, which separates the enamel from the internal substance," or dentine.

The calcification and conversion of the cells of the dentinal pulp commence as usual at the peripheral parts of the lamelliform processes furthest from the attached base. It may readily be conceived, therefore, that, at the commencement, there is formed a little cap upon each of the processes into which the edges of the pulp-plates are divided. As the centripetal calcification proceeds, the caps are converted into horn-shaped cones. When it has reached the bottom of the notches of the edge of the pulp-plate, all the cones become united together into a single transverse plate; and, the process of conversion, having reached the base of the pulp-plate, these plates coalesce to form a common base to the crown of the tooth, which would then present the same eminences and notches that characterized the gelatinous pulp, if, during the period of conversion, other substances had not been formed upon the surface and in the interspaces of the pulp-plates.

Coincident, however, with the formation of the dentine, is the deposition of the hardening salts of the enamel in the extremely slender prismatic cells, which are for the most part vertical to the plane of the inner surface of the folds of the capsule to which they are attached. The true inner part of the capsule forms those thick transverse folds or partitions which support the enamel organ, and with it fill the interspaces of the dentinal pulps. With regard to the formation of the cement, Cuvier, after citing the opinion of Tenon—that it was the result of ossification of the internal layer of the capsule, and that of Blake—that it was a deposition from the opposite surface of the capsule to that which had deposited the enamel, states his own conviction to be that the cement is produced by the same layer and by the same surface as that which has produced the enamel. The proof alleged is, that so long as any space remains between the cement and the external capsule, that space is found to contain a soft internal layer of the capsule with a free surface next the cement. The phenomena could not, in fact, be otherwise explained according to the "excretion theory" of dental development. To the obvious objection that the same part is made, in this explanation, to secrete two different products, Cuvier replies, that it undergoes a change of tissue: "Whilst it yielded enamel only, it was thin and transparent; to give cement, it becomes thick, spongy, and of a reddish color." The external characters of the enamel-organ and cement-forming capsule are correctly defined; only, the one, instead of being converted into the other, is in fact changed into its supposed transudation; the enamel fibres being formed, and properly disposed in the direction in which their chief strength is to lie, by the assimilative properties of the prearranged elongated, prismatic, non-nucleated cells, which take from the surrounding plasma the required salts, and compact them in their interior.

Whilst this process is on foot, and before the enamel fibres are firm in their position, the capsule begins to undergo that change which results in the formation of the thick cement; the calcifying process commences from several points, and proceeds centrifugally, radiating therefrom, and differing from the ossification of bone chiefly in the number of these centres, which, though close to the new-formed enamel, are in the substance of the inner vascular surface of the capsular folds. The cells arrange themselves in concentric layers around the vessels, and act like those of the enamel pulp in receiving into their interior the bone-salts in a clear and compact state. During this process they become confluent with each other, their primitive distinctness being indicated only by their persistent granular nuclei, which now form the radiated Purkingian capsules. The interspaces of the concentric series of confluent cells become filled with the calcareous salts in a rather more opaque state; and the conversion of the capsule into cement goes on, according to the processes more particularly described in the Introduction to my *Odontography*, until a continuous stratum is formed in close connection with the layer of enamel.

Calcification extending from the numerous centres, the different portions coalesce, and progressively add to the thickness of the cement, until all the interspaces of the coronal plates and the whole exterior of the crown are covered with the bone-like substance. The enamel-pulp ceases to be developed at the base of the crown, but the capsule continues to be formed *pari passu* with the partial formation of the pulp, as this continues, progressively contracting, from the base of the crown, to form, by its calcification, the roots. The calcification of the capsule going on at the same time, a layer of cement is formed in immediate connection with the dentine. The circumscribed spaces at the bottom of the socket to which the capsule and dentinal pulp adhere, where they receive their vessels and nerves, and which are the seat of the progressive formation of these respective moulds of the two dental tissues, become gradually contracted, and subdivided by the further localization of the reproductive forces to particular spots, whence the subdivision of the base into roots. The surrounding bone undergoes corresponding modifications, growing and filling up the interspaces left by the dividing and contracting points of attachment of the residuary matrix. All is subordinated to one harmonious law of growth by vascular action and cell-formation, and of molecular decrement by absorption. Mechanical squeezing, or drawing out, has no share in these changes of the pulp or capsule; pressure at most exercises only a gentle stimulus to the vital processes. Cuvier believed that there were places where the dentinal pulp and the capsule were separate from each other. I have never found such, except where the enamel-pulp was interposed between them in the crown of the tooth, or where both pulp and capsule adhered to the periosteum of the socket, below the crown. Cuvier affirms that the number of fangs of an elephant's molar depends upon the number of points at which the base of the gelatinous (dentinal) pulp is attached to the bottom of the capsule; and that the interspaces of these attachments constitute the under part of the crown or body of the tooth, the attachments themselves forming the first beginnings of the fangs. True to his hypothesis of the formation of the dental tissues by excretion, he says that the elongation of the fangs is produced by two circumstances; first, the progressive elongation of the layers of osseous substance (dentine) which force the tooth to rise and emerge from its socket; secondly, the thickening of the body of the tooth by the addition of successive layers to its inner surface, which, filling up the interior cavity, leaves scarcely room for the gelatinous pulp, and forces it down into the interior of the roots.

This pulling up of the fang on the one hand, and squeezing down the pulp on the other, are forces too gross and mechanical to be admitted in actual physiology to explain the growth of the root of a tooth or of any other organized product; such modes of explanation were, however, inevitable in adopting the "excretion theory" of dental development.

There are few examples of organs that manifest a more striking adaptation of a highly complex and beautiful structure to the exigences of the animal endowed with it, than the grinding teeth of the elephant. We perceive, for example, that the jaw is not incumbered with the whole weight of the massive tooth at once, but that it is formed by degrees as it is required; the division of the crown into a number of successive plates, and the subdivision of these into cylindrical processes, presenting the conditions most favorable to progressive formation. But a more important advantage is gained by this subdivision of the tooth; each part is formed like a perfect simple tooth, having a body of dentine, a coat of enamel, and an outer investment of cement. A single digital process may be compared to the simple canine of a carnivore; a transverse row of these, therefore, when the work of mastication has commenced, presents by virtue of the different densities of their constituent substances, a series of cylindrical ridges of enamel, with as many depressions of dentine, and deeper external valleys of cement; the more advanced and more abraded part of the crown is traversed by the transverse ridges of the enamel inclosing the depressed surface of the dentine, and separated by the deeper channels of cement; the forepart of the tooth exhibits its least efficient condition for mastication, the inequalities of the grinding surface being reduced, in proportion as the enamel and cement have been worn away. This part of the tooth is, however, still fitted for the first coarse crushing of the branches of a tree; the transverse enamel ridges of the succeeding part of the tooth divide it into smaller fragments, and the posterior islands and tubercles of enamel pound it to the pulp fit for deglutition.

The structure and progressive development of the tooth not only give to the elephant's grinder the advantage of the uneven surface which adapts the millstone for its office, but, at the same time, secure the constant presence of the most efficient arrangement for the finer comminution of the food, at the part of the mouth which is nearest the fauces.

In the tusks of the *Mastodon giganteus*, the outer layer of cement is relatively thicker than in the tusks of the mammoth, or in those of the Indian elephant. The general character of the microscopic structure of the ivory of the mastodon's tusk is the same as that of the elephant.

By the minuteness and close arrangement of the dentinal tubes, and especially by their strongly undulating secondary curves, a tougher and more elastic tissue is produced than results from their disposition in ordinary dentine; and the modification which distinguishes "ivory" is doubtless essential to the due degree of coherence of so large a mass as the elephant's tusk, projecting so far from the supporting socket; and to be frequently applied in dealing hard blows and thrusts.

TEETH OF THE MEGATHERIUM.

The megatherium (Gr. *megas*, great; *therion*, beast), so called from its colossal size—being as large as the elephant, and even surpassing that hugest of existing quadrupeds in some of its proportions—was once an inhabitant, and apparently in some numbers, of the American continent, especially its southern division, and subsisted on a similar kind of food to the elephants, viz: the smaller branches and leaves of trees; but all the genera and species of megatheroid beasts are now extinct. Nevertheless, from the fossil remains of the megatherium the anatomist is able unerringly to deduce the nature of its food and many of its peculiar habits; and also to bring to light a system of dentition, designed, like that of the elephants, for the service of crushing and masticating a coarse vegetable diet throughout a long-protracted individual existence; and yet, by a modification of the formative processes and economy of the teeth, quite different from those that have been adopted for the same ends in the elephant tribe.

In these, as has been shown, the supply of a masticating apparatus, to serve the requirements of a gigantic animal during one or perhaps two centuries of existence, was provided by a succession of *different* molar teeth presenting the due complexity of structure. In the megatherium, the same end was obtained by a perpetual growth of the *same* complex molar teeth—the different dental substances being formed at and added to the base of the tooth, in proportion as they were ground down at the exposed summit.

The true number of teeth was determined by a removal of the mineral substances adhering to the surface of a portion of a fossil skull of a megatherium, brought by Mr. Charles Darwin, from South America (Fossil Mammalia of the "Voyage of the Beagle," 4to. 1840, p. 102). The animal has not, as in the elephant, any tusks; its teeth are molars or grinders exclusively; they are five in number on each side of the upper jaw, and four on each side of the lower jaw—eighteen in all. All these teeth are remarkable for their great length in proportion to their breadth or thickness, being from eight to ten inches in length, and between two and three inches only in breadth. They are very deeply implanted in the jaw, and the lower jaw has a quite peculiar form, in order to acquire the requisite room for the lodgement of the lower teeth and their "matrices," or formative organs.

The next peculiarity to be noticed in these remarkable teeth is the great length of the conical cavity at their base, for lodging the part of the matrix called the "pulp;" the apex of the pulp-cavity rising as far as the part of the tooth where it emerges from the socket. A transverse fissure is continued from this apex to the middle concavity of the grinding surface of the tooth, which is thus divided into two halves. Each of these halves consists of three distinct substances—a central column of "vaso-dentine," a peripheral and nearly equally thick layer of "cement," and an intermediate thinner stratum of true or "hard dentine." This latter has been described as being enamel; but it is only analogous to that differently constituted and harder substance in the compound teeth of the elephant, in regard to its relative situation, and its degree of density to the other constituents of the tooth of the megatherium.

No species of the order called "Bruta," or "Edentata," to which the extinct megatherium belongs, has true enamel entering into the composition of its teeth; but the modifications of structure which the teeth present in the different genera of this order are considerable, and their complexity is not less that that of the enamelled teeth of the herbivorous, ruminant, and other hoofed animals, in consequence of the introduction of a dental substance—the "vaso-dentine"—into their composition, analogous in structure to that of the teeth of the *Myliobates* and other cartilaginous fishes. The cement of the megatherium's tooth differs from the vaso-dentine in the larger size and wider interspaces of its medullary canals, and by the presence of radiated bone-cells in their interspaces; but they are brought into organic communication with each other, not only by means of the tubes of coarse dentine, but by occasional continuity of the vascular canals across that substance. The tooth of the megatherium thus offers an unequivocal example of a course of nutriment from the dentine to the cement, and reciprocally; so that the main substance, or body of the tooth, can obtain the requisite supply for its languid vitality from the vessels of the capsule as well as from those of the pulp.

The conical cavity at the base of the tooth attests the large size, and demonstrates the form of the persistent pulp in the living megatherium; the diameter of its base is equal to the part of the tooth which is formed by the combined dentine and vaso-dentine. From the gradual thinning off and final disappearance of those substances as they reach the base of the tooth, it may be inferred that both were formed at the expense of the pulp. The fine dentinal tubes must have been established and calcified in the peripheral layer of the pulp, which layer must have been wholly so converted into the dentine; but as the deposition of the hardening salts proceeded in the rest of the pulp, certain tracts of that soft and vascular substance were left uncalcified, to form the medullary or vascular canals which characterize the vaso-dentine. The space between the inserted base of the tooth and the walls of the socket indicates the thickness of the dental capsule, by the ossification of which the exterior layer of cement was formed; and this modification of the tooth-forming organ in the megatherium permitted the progressive addition of cement, as the persistence of the compound pulp occasioned the uninterrupted and continuous formation of the harder dentine, which is analogous to the enamel in the elephant's grinder.

In all essential characters the teeth of the megatherium repeat, on a magnified scale, the dental peculiarities of the sloth; and since, from a similarity of the form, number, kinds, and structure of teeth, a similarity of food is to be inferred, it may be concluded that the leaves and soft succulent sprouts of trees formed the staple diet of the megatherium, and of the cognate and contemporary megalonyx and mylodon, as of the existing sloths. The enormous claws of those great extinct sloth-like quadrupeds, to judge by the fossorial (digging and scratching) character of the powerful mechanism of the limbs that worked them, were employed, not, as in the sloths, to carry the animal to its food, but to bring the food within the reach of the animal, by uprooting the trees on which it grew.

In the remains of the megatherium, we have evidence of the framework of a quadruped equal to the task of undermining and tearing down the largest trees in a tropical forest. In the latter operation, it is obvious that the immediate application of the anterior extremities to the trunk of the tree would demand a corresponding fulcrum to be effectual; and it is the necessity for an adequate basis of support and resistance to such an application of the fore-extremities which gives the explanation of the seemingly anomalous development of the pelvis, tail, and hinder extremities of the megatherium and its extinct allies. No wonder, therefore, that their type of structure should be so peculiar; for, where shall we now find quadrupeds equal, like them, to the habitual task of uprooting trees for food.

TEETH OF THE ANOPLOTHERIUM.

Of the extinct quadrupeds with hoofs, and which were consequently herbivorous, the species restored by Cuvier from fossil remains discovered in the quarries at Montmartre, near Paris, was one of the most ancient. The great comparative anatomist called it *anoplotherium*, from the Greek words signifying "weaponless," because it had neither horns nor tusks. It was, however, characterized by the most complete system of dentition; for it not only possessed incisors and canines in both jaws, but these were so equably developed that they formed one unbroken series with the premolars and molars, which character is now found only in the human species.

The dental formula of the genus *Anoplotherium* is expressed by: $i\,\frac{3-3}{3-3}, c\,\frac{1-1}{1-1}, p\,\frac{4-4}{4-4}, m\,\frac{3-3}{3-3} = 44$, signifying that it had, on each side of both upper and lower jaws, three incisors, one canine, four premolars, and three true molars; in all, forty-four teeth.

Those teeth which are transitorily manifested in the embryo state of some ruminants, as the upper incisors and canines and the anterior premolars, $p\,1$, were in the ancient anoplothere retained and raised to a proportional equality of size and function with the rest of the teeth. The true molars had a broad grinding surface, with enamel-covered crescentic lobes, remotely resembling those of the existing ruminants. In some of the smaller species of *anoplotherium*, the ruminant type of grinding surface was more closely adhered to, and the fossil lower jaws of such species, as *e.g.* of the *Dichobune cervinum*, have been mistaken for those of a ruminant, and have been referred to the genus *Moschus*. One of these interesting transitional extinct quadrupeds, described in the *Geological Journal*, for 1847, under the name of *Dichodon*, had forty-four teeth in one uninterrupted series, and of the same kinds, as in the anoplothere; but the teeth there marked *p* 4, and *m* 1, upper jaw, I have ascertained to be "milk-teeth."

TEETH OF RUMINANTS.

The even-toed or artiodactyle *Ungulata*, superadd the characters of simplified form and diminished size to the more important and constant one of vertical succession in their premolar teeth. These teeth, in the ruminants, represent only the moiety of the true molars, or one of the two semi-cylindrical lobes of which these teeth consist, with at most a rudiment of the second lobe. An analogous morphological character of the premolars will be found to distinguish them in the dentition of the genus *Sus* (Fig. 75, *p* 2, *p* 3, *p* 4), in the hippopotamus, and in the *phacochœrus*, or wart-hog, where the premolar series is greatly reduced in number; yet this instance of a natural affinity, manifested in so many other parts of the organization of the artiodactyle genera, has been overlooked in F. Cuvier's work, above cited, although it is expressly designed to show how much zoological relations are illustrated by the teeth.

Most of the deciduous teeth of the ruminants resemble in form the true molars; the last, *e. g.* has three lobes in the lower jaw like the last true molar. When, therefore, the third grinder of the lower jaw of any new or rare ruminant shows three lobes, the crowns of the premolars should be sought for in the substance of the jaw below these, and above their opponents in the upper jaw; and thus the true characters of the permanent dentition may be ascertained.

The deciduous molars are three in number on each side, and, being succeeded by as many premolars, the ordinary permanent molar formula is $p\ \frac{3-3}{3-3},\ m\ \frac{3-3}{3-3}$: but there is a rudiment of an anterior milk-molar, $d\,1$, in the embryo fallow-deer, and in one of the most ancient of the extinct ruminants (*dorcatherium*, Kaup) the normal number of premolars was fully developed.

The molar series of all the Diphyodonts is naturally divisible into only two groups, premolars and molars; the typical number of these is $\frac{4-4}{4-4}, \frac{3-3}{3-3}$; and each individual tooth may be determined and symbolized throughout the series, as is shown in the instances under Cut 75.

SEAL TRIBE.

(*Phocidæ*).—There is a tendency to deviate from the ferine number of the incisors in the most aquatic and piscivorous of the Musteline quadrupeds, viz., the sea-otter (*enhydra*), in which species the two middle incisors of the lower jaw are not developed in the permanent dentition. In the family of true seals, the incisive formula is further reduced, in some species even to zero in the lower jaw, and it never exceeds $\frac{3-3}{2-2}$. All the *phocidæ* possess powerful canines; only in the aberrant walrus are they absent in the lower jaw; but this is compensated by the singular excess of development which they manifest in the upper jaw (Fig. 70). In the pinnigrade, as in the plantigrade family of carnivores, we find the teeth which correspond to true molars more numerous than in the digitigrade species, and even occasionally rising to the typical number, three on each side; but this, in the seals, is manifested in the upper, and not, as in the bears, in the lower jaw. The entire molar series usually includes five, rarely six teeth on each side of the upper jaw, and five on each side of the lower jaw, with crowns, which vary little in size or form in the same individual; they are supported in some genera, as the eared seals (*otariæ*), and elephant seals (*cystophora*), by a single fang; in other genera by two fangs, which are usually connate in first or second teeth; the fang or fangs of both incisors, canines and molars, are always remarkable for their thickness, which commonly surpasses the longest diameter of the crown. The crowns are most commonly compressed, conical, more or less pointed; in a few of the largest species they are simple and obtuse, and particularly so in the walrus, in which the molar teeth are reduced to a smaller number than in the true seals. In these, the line of demarcation between the true and false molars is very indefinitely indicated by characters of form or position; but, according to the instances in which a deciduous dentition has been observed, the first three permanent molars in both jaws succeed and displace the same number of milk molars, and are consequently *premolars;* occasionally, in the seals with two-rooted molars, the more simple character of the premolar teeth is manifested by their fangs being connate, and in the *Stonorhynchus serridens* the more complex character of of the true molars is manifested in the crown. In the *Stenorhynchus leptonyx*, each molar tooth in both jaws is trilobed, the anterior and posterior accessory curving towards the principal one, which is bent slightly backwards; all the divisions are sharp-pointed, and the crown of each molar thus resembles the trident or fishing-spear; the two fangs of the first molar in both jaws are connate. In the *Stenorhyncus serridens*, the three anterior molars on each side of both jaws are four-lobed, there being one anterior and two posterior accessory lobes; the remaining posterior molars (true molars) are five-lobed, the principal cusp having one small lobe in front, and three developed from its posterior margin; the summits of the lobes are obtuse, and the posterior ones are recurved like the principal lobe. Sometimes the third molar below has three instead of two posterior accessory lobes. Occasionally, also, the second, as well as the first molar above, has its fangs connate: but the essentially duplex nature of the seemingly single fang, which is unfailingly manifested within by the double pulp-cavity, is always outwardly indicated by the median longitudinal opposite indentations of the implanted base.

TEETH OF QUADRUMANA.

The chief aim of comparative anatomy being the better comprehension of the structure of man, we shall finally describe those modifications of the dental system which throw more immediate light on the nature of the teeth in the human subject, and which are met with, as might be expected, in the order (*Quadrumana*) of *mam-*

malia that makes the nearest approach to that represented by the genus *homo.*

Through a considerable part of the quadrumanous series, *e.g.*, in all the apes and monkeys of the Old World, in all the genera, indeed, which are above the lemurs (cat-monkeys and slow monkeys) of Madagascar, the same number and kinds of teeth are present as in man; the first deviation being the disproportionate size of the canines and the concomitant break, or "diastema," in the dental series for the reception of their crowns when the mouth is shut. This is manifested in both the chimpanzees and orangs, together with a sexual difference in the proportions of the canine teeth.

In that large ape of tropical Africa, called the "gorilla" (*Troglodytes gorilla*), which, in some important particulars, more resembles man than does the smaller kind of chimpanzee (*Troglodytes niger*), the dentition seems to approach nearer to the carnivorous type, at least in the full-grown male (see Fig. 50, p. 223). It is nevertheless strictly quadrumanous in its essential characters, as in the broad, flat, tuberculate grinding surfaces of the molar teeth; but in the minor particulars in which it differs from the dentition of the orang, it approaches nearer the human type. In the upper jaw the middle incisors are smaller, the lateral ones larger than those of the orang; they are thus more nearly equal to each other; nevertheless, the proportional superiority of the middle pair is much greater than in man, and the proportional size of the four incisors both to the entire skull and to the other teeth is greater. Each incisor has a prominent posterior basal ridge, and the outer angle of the lateral incisors, *i* 2, is rounded off as in the orang. The incisors incline forwards from the vertical line as much as in the great orang. The characteristics of the human incisors are, in addition to their true incisive wedge-like form, their near equality of size, their vertical or nearly vertical position, and small relative size to the other teeth and to the entire skull. The diastema, between the incisors and the canine on each side, is as well marked in the male chimpanzee as in the male orang. The crown of the canine (*ib.*), *c*, passing outside the interspace between the lower canine and premolar, extends, in the male *Troglodytes gorilla*, a little below the alveolar border of the under jaw when the mouth is shut: the canines in both jaws are twice the size of those teeth in the female gorilla.

Both premolars are bicuspid; the outer cusp of the first and the inner cusp of the second being the largest, and the first premolar consequently appearing the largest on an external view. The anterior external angle of the first premolar is not produced as in the orang, which, in this respect, makes a marked approach to the lower *quadrumana*. In man, where the outer curve of the premolar part of the dental series is greater than the inner one, the outer cusps of both premolars are the largest; the alternating superiority of size in the chimpanzee accords with the straight line which the canine and premolars form with the true molars.

The three true molars are quadricuspid, relatively larger in comparison with the bicuspids than in the orang. In the first and second molars of both species of chimpanzee, a low ridge connects the antero-internal with the postero-external cusp, crossing the crown obliquely, as in man. There is a feeble indication of the same ridge in the unworn molars of the orang; but the four principal cusps are much less distinct, and the whole grinding surface is flatter and more wrinkled than in the chimpanzee. The repetition of the strong sigmoid curves, which the unworn prominences of the first and second true molars present in man, is a very significant indication of the near affinity of the gorilla and the chimpanzee, as compared with the approach made by the orangs or any of the inferior *quadrumana*, in which the four cusps of the true molars rise distinct and independently of each other. The premolars as well as molars are severally implanted by one internal and two external fangs, diverging, but curving towards each other at their ends as if grasping the substance of the jaw. In no variety of the human species are the premolars normally implanted by three fangs; at most the root is bifid, and the outer and inner divisions of the root are commonly connate. It is only in the black varieties, and more particularly that race inhabiting Australia, that I have found the wisdom tooth, or last true molar, with three fangs as a general rule; and the two outer ones are more or less confluent.

The molar series in both species of chimpanzee forms a straight line, with a slight tendency in the upper jaw to bend in the opposite direction to the well-marked curve which the same series describes in the human subject. This difference of arrangement, with the more complex implantation of the premolars, the proportionally larger size of the incisors as compared with the molars; the still greater relative magnitude of the canines; and, above all, the sexual distinction in that respect illustrated by the skull of the full-grown male gorilla (Fig. 50, p. 223), stamp the chimpanzees most decisively with not merely specific but generic distinctive characters as compared with man. For the teeth are fashioned in their shape and proportions in the dark recesses of their closed formative alveoli, and do not come into the sphere of operation of external modifying causes, until the full size of the crowns has been acquired. The formidable natural weapons, with which the Creator has armed the powerful males of both species of chimpanzee, form the compensation for the want of that psychical capacity to forge destructive instruments which has been reserved as the exclusive prerogative of man. Both chimpanzees and orangs differ from the human subject in the order of the development of the permanent series of teeth; the second molar, *m* 2, comes into place before either of the premolars has cut the gum, and the last molar, *m* 3, is acquired before the canine. We may well suppose that the larger grinders are earlier required by the frugivorous chimpanzees and orangs than by the higher organized omnivorous species with more numerous and varied resources; and probably one main condition of the earlier development of the canines and premolars in man may be their smaller relative size.

In the South American quadrumana, the number of teeth is increased to thirty-six, by an addition of one tooth to the molar series on each side of both jaws. It might be concluded, *à priori*, that as three is the typical number of true molars in the placental mammalia with two sets of teeth, the additional tooth in the *cebinæ* would be a premolar, and form one step to the resumption of the normal number (four) of that kind of teeth. The proof of the accuracy of this inference is given by the state of the dentition in any young spider-monkey (*Ateles*), or Capucin-monkey (*Cebus*), which may correspond with that of the human child in Fig. 76, *i. e.* where the whole of the deciduous dentition is retained, together with the first true molar (*m* 1) on each side of both jaws. If the germs of the other teeth of the permanent series be exposed in the upper jaw (as in Fig. 76), the crown of a premolar will be found above the third molar in place, as well as above the second and first. As regards number, therefore, the molar series, in the South American monkeys (*Mycetes*, *Ateles*, *Cebus*), is intermediate between that of the genus *Mustela* and of *Felis* (Fig. 69); the little premolar, *p i*, in *Mustela*, shows plainly enough which of the four is wanting to complete the typical number in the South American monkey, and which is the additional premolar distinguishing its dental formula from that of the Old World monkeys and man.

Zoologists have rightly stated, as a matter of fact, that the little marmoset monkeys (*Hapale*, *Ouistiti*) "have only the same number of teeth as the monkeys of the Old World—viz: 32, $i\frac{4}{4}, c\frac{1-1}{1-1}, m\frac{5-5}{5-5}$." But the difference is much greater than this numerical conformity would intimate. In a young *Jacchus penicillatus*, I find that there are three deciduous molars displaced by three premolars, as in the other South American quadrumana, and that it is the last true molar, *m* 3, the development of which is suppressed, not the premolar, *p* 2, and thus these diminutive squirrel-like monkeys actually differ from the Old World forms more than the *Cebidæ* do; *i. e.* they differ not only in having four teeth ($p\ 2\ \frac{1-1}{1-1}$), which the monkeys of the Old World do not possess, but also by wanting four teeth ($m\ 3\ \frac{1-1}{1-1}$), which those monkeys, as well as the *Cebidæ*, actually have. It is thus that the investigation of the exact homologies of parts leads to a recognition of the true characters indicative of zoological affinity.

Most of the *Lemurinæ* have $p\frac{3-3}{3-3}, m\frac{3-3}{3-3}$, together with remarkable modifications of their incisive and canine teeth, of which an extreme example is shown in the pectinated tooth of the *galeopithecus*. The inferior incisors slope forwards in all, and the canines also, which are contiguous to them, and very similar in shape.

In the hoofed quadrupeds with toes in uneven number (*perissodactyla*), whose premolars, for the most part, repeat both the form and the complex structure of the true molars, such premolars are distinguished by the same character of development as those of the *artiodactyla*, or ungulates, with toes in even number; although here the premolars are distinguished also by modifications of size and shape. The complex ridged and tuberculate crowns of the second, third, and fourth grinders of the rhinoceros, hyrax, and horse, no more prove them to be true molars than the trenchant shape of the lower carnassials of the lion proves them to be false molars. It is by development alone that the primary division of the series of grinding teeth can be established, and by that character only can the homologies of each individual tooth be determined, and its proper symbol applied to it.

In Fig. 72, the three posterior teeth of the almost uniform grinding series of the horse's dentition are thus proved to be the only ones entitled to the name of "true molars;" and, if any one should doubt the certainty of the rule of counting, by which the symbols, *p* 4, *p* 3, and *p* 2, are applied to the three large anterior grinding teeth (*ib.*), which are commonly the only premolars present in each lateral series of the horse's jaws, yet the occasional retention of the diminutive tooth (*p* 1), would establish its accuracy, whether such tooth be regarded as the first of the deciduous series unusually long retained, or the unusually small and speedily lost successor (*p* 1) of an abortive (*d* 1).

The law of development, so beautiful for its instructiveness and constancy in the placental *diphyodonts*, is well illustrated in the little hyrax, in which the *d* 1 is normally developed and succeeded by a permanent, *p* 1, differing from the rest only by a graduated inferiority of size, which, in regard to the last premolar, ceases to be a distinction between it and the first true molar.

The elephant, which by its digital characters belongs to the odd-toed, or perissodactyle, group of pachyderms, also resembles them in the close agreement in form and structure of the grinding teeth representing the premolars, with those that answer to the true molars of the hyrax, tapir, and rhinoceros. The gigantic proboscidian pachyderms of Asia and Africa present, however, so many peculiarities of structure as to have led to their being located in a particular family in the Systematic Mammalogies. And this seems to be justified by no character more than by the singular seeming exception which they present to the diphyodont rule which governs the dentition of other hoofed quadrupeds. In fact, the elephant, like the dugong, sheds and replaces vertically only its incisors, which are also two in number, very long, and of constant growth, forming tusks, with an analogous sexual difference in this respect in the female of the Asiatic species. The molars, also, are successively lost, are not vertically replaced, and are reduced finally to one on each side of both jaws, which is larger than any of its predecessors. These analogies are interesting and suggestive in connection with the other approximations in the "Sirenia" to the pachydermal type.

In the mammalian orders with two sets of teeth, these organs acquire fixed individual characters, receive special denominations, and can be determined from species to species. This individualization of the teeth is eminently significative of the high grade of organization of the animals manifesting it. Originally, indeed, the name "incisors," "laniaries," or "canines," and "molars" were given to the teeth, in man and certain mammals, as in reptiles, in reference merely to the shape and offices so indicated; but they are now used as arbitrary signs, in a more fixed and determinate sense. In some carnivora, *e. g.* the front teeth have broad tuberculate summits, adapted for nipping and bruising, while the principal back teeth are shaped for cutting, and work upon each other like the blades of scissors. The front teeth in the elephant project from the upper jaw, in the form, size, and direction of long-pointed horns. In short, shape and size are the least constant of dental characters in the mammalia; and the homologous teeth are determined, like other parts, by their relative position, by their connections, and by their development.

Those teeth which are implanted in the premaxillary bones, and in the corresponding part of the lower jaw, are called "incisors," whatever be their shape or size. The tooth in the maxillary bone, which is situated at, or near to, the suture with the premaxillary, is the "canine," as is also that tooth in the lower jaw which, in opposing it, passes in front of its crown when the mouth is closed. The other teeth of the first set are the "deciduous molars;" the teeth which displace and succeed them vertically are the "premolars;" the more posterior teeth, which are not displaced by vertical successors, are the "molars" properly so called.

The hog is one of the few existing quadrupeds which retain the typical number and kinds of teeth.

Figure 75, part of the lower jaw of a young hog, illustrates the phenomena of development which distinguishes the premolars from the molars. The first premolar, *p* 1,

Fig. 75.

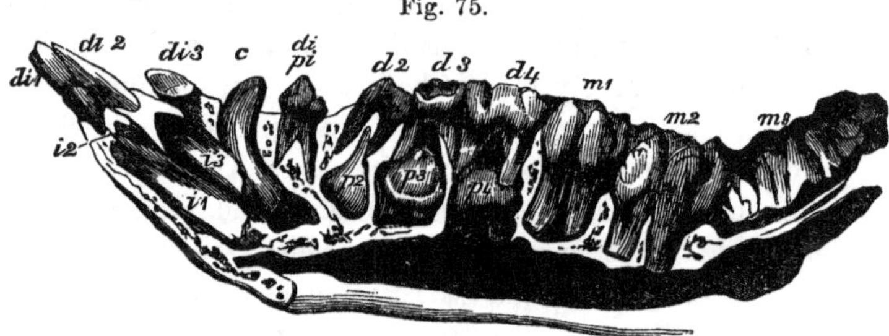

DECIDUOUS AND PERMANENT TEETH OF THE HOG.

and the first molar, *m* 1, are in place and use, together with the three deciduous molars, *d* 2, *d* 3, and *d* 4; the second molar, *m* 2, has just begun to cut the gum; *p* 2, *p* 3, and *p* 4, together with *m* 3, are more or less incomplete, and concealed in their closed alveoli.

The premolars must displace deciduous molars in order to rise into place; the molars have no such relations. It will be observed that the last deciduous molar, *d* 4, has the same relative superiority of size to *d* 3, and *d* 2, which *m* 3 bears to *m* 2 and *m* 1; and the crowns of *p* 3 and *p* 4 are of a more simple form than those of the milk-teeth which they are destined to succeed.

The germ of the permanent canine has not yet appeared below the deciduous one, *c*; those of the permanent incisors, *i* 1, *i* 2, *i* 3, are seen ready to push out the deciduous incisors *d* 1, *d* 2, *d* 3. When the whole of the second set of teeth is in place, its nature is indicated by the formula:—$i\frac{3-3}{3-3}, c\frac{1-1}{1-1}, p\frac{4-4}{4-4}, m\frac{3-3}{3-3} = 44$: which signifies that there are, on each side of both upper and lower jaws, three incisors, one canine, four premolars, and three molars, making in all forty-four teeth; each distinguished by the symbol marked in the cut.

When the premolars and the molars are below their typical number, the absent teeth are missing from the fore part of the premolar series, and from the back part of the molar series. The most constant teeth are the fourth premolar and the first true molar; and, these being known by their order and mode of development, the homologies of the remaining molars and premolars are determined by counting the molars from before backwards, *e. g.*, "one," "two," "three," and the premolars from behind forwards, "four," "three," "two," "one."

Examples of the typical dentition are exceptions in the actual creation; but it was the rule in the forms of mammalia first introduced into this planet; and that, too, whether the teeth were modified for animal or vegetable food.

With regard to the human dentition, the discovery by the great poet Goethe of the limits of the premaxillary bone in man, leads to the determination of the incisors, which are reduced to two on each side of both jaws; the contiguous tooth shows by its shape, as well as position, that it is the canine, and the characters of size and shape have also served to divide the remaining five teeth in each lateral series into two bicuspids and three molars. In this instance, the secondary characters conform with the essential ones. But, since we have seen of how little value shape or size are, in the order carnivora, in the determination of the exact homologies of the teeth, it is satisfactory to know that the more constant and important character of development gives the requisite certitude as to the nature of the so-called bicuspids in the human subject. In Fig. 76, the condition of the teeth is shown in

Fig. 76.

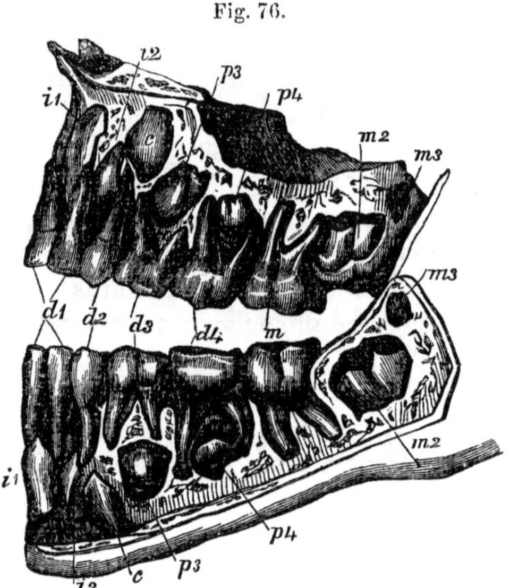

DECIDUOUS AND PERMANENT TEETH, HUMAN, ÆT. 7.

the jaws of a child of about six years of age. The two incisors on each side, *d*, *i*, are followed by a canine, *c*, and this by three molar teeth like those of the adult; in fact, the last of the three, *m*, is the first of the permanent molars; it has pushed through the gum, like the two molars which are in advance of it, without displacing any previous tooth, and the substance of the jaw contains no germ of any tooth destined to displace it; it is, therefore, by this character of its development, a true molar, and the germs of the permanent teeth, which are exposed in the substance of the jaw, between the diverging fangs of the molars *d* 3 and *d* 4, prove those molars to be temporary, destined to be replaced, and prove also that the teeth about to displace them are premolars. According, therefore, to the rule previously laid down, we count the permanent molar in place as the first of its series, *m* 1, and the adjoining premolar as the last of its series, and consequently the fourth of the typical dentition, or *p* 4.

We are thus enabled, with the same scientific certainty as that whereby we recognize in the middle toe of the foot the homologue of that great digit which forms the whole foot and is incased by the hoof of the horse, to point to *p* 4, or the second bicuspid in the upper jaw, and to *m* 1, or the first molar in the lower jaw, of man, as the homologues of the great carnassial, or flesh-cutting teeth of the lion (Fig. 69). We also conclude that the teeth which are wanting in man to complete the typical molar series are the first and second premolars, the homologues of those marked *p* 1 and *p* 2 in the hog. The characteristic shortening of the maxillary bones required this diminution of the number of their teeth, as well as their size, and of the canines more especially; and the still greater curtailment of the premaxillary bone is attended with a diminished number and an altered position of the incisors.

The homologous teeth being thus determinable, they may be severally signified by a symbol as well as by a name. The incisors, *e. g.*, are here represented by their initial letter, *i*, and individually by an added number, *i* 1, *i* 2, and *i* 3; the canines by the letter *c;* the premolars by the letter *p;* and the molars by the letter *m;* these also being differentiated by added numerals. Thus the number of these teeth, on each side of both jaws, in any given species—man, *e. g.*—may be expressed by the following brief formula: $i\frac{2-2}{2-2}$, $c\frac{1-1}{1-1}$, $p\frac{2-2}{2-2}$, $m\frac{3-3}{3-3}=32$; and the homologies of the typical formula may be signified by *i* 1, *i* 2; *c*; *p* 3, *p* 4; *m* 1, *m* 2, *m* 3; the suppressed teeth being *i* 3, *p* 1, and *p* 2.

These symbols, it is hoped, are so plain and simple as to have formed no obstacle to the full and easy comprehension of the facts explained by means of them. If these facts, in the manifold diversities of mammalian dentition, were to be described in the ordinary way, by means of verbal phrases or definitions of the teeth—*e. g.* "the second deciduous molar, representing the fourth in the typical dentition," instead of *d* 4, and so on—the description would occupy much space, and would levy such a tax upon the attention and memory as must tend to enfeeble the judgment, and impair the power of seizing and appreciating the results of the comparisons.

Each year's experience strengthens my conviction that the rapid and successful progress of the knowledge of animal structures, and of the generalization deducible therefrom, will be mainly influenced by the determination of the nature or homology of the parts, and by the concomitant power of condensing the propositions relating to them, and of attaching to them signs or symbols equivalent to their single substantive names. In my work on the *Archetype of the Skeleton*, I have denoted most of the bones by simple numerals, which, if generally adopted, might take the place of names; and all the propositions respecting the centrum of the occipital vertebra might be predicated of the figure "1" as intelligibly as of "basi-occipital."

The symbols of the teeth are fewer, are easily understood and remembered, render unnecessary the endless repetition of the verbal definitions of the parts, harmonize conflicting synonymes, serve as a universal language, and express the author's meaning in the fewest and clearest terms. The entomologist has long found the advantage of such signs as ♂ and ♀, signifying male and female, and the like; and it is time that the anatomist should avail himself of this powerful instrument of thought, instruction, and discovery, from which the chemist, the astronomer, and the mathematician have obtained such important results.

INDEX,

GLOSSARIAL, EXPLANATORY, AND REFERENTIAL.

A

B

F

M

N

V

W

X

Z

THE END.

CATALOGUE

OF

MEDICAL, SURGICAL, AND SCIENTIFIC WORKS,

PUBLISHED BY

BLANCHARD & LEA, PHILADELPHIA.

AMERICAN JOURNAL OF THE MEDICAL SCIENCES.—Edited by ISAAC HAYS, M.D. Published Quarterly, each number containing about 300 large octavo pages. Price, $5 per annum. When paid for in advance, it is sent free by post, and the "Medical News and Library," a monthly of 32 large 8vo. pages, is furnished gratis. Price of the "Medical News," separate, $1 per annum, in advance.

ALLEN (J. M.), M.D.—THE PRACTICAL ANATOMIST; OR, THE STUDENT'S GUIDE IN THE DISSECTING-ROOM. With over 200 illustrations. In one handsome royal 12mo. volume. (*Just Issued.*)

ABEL (F. A.), F.C.S., and C. L. BLOXAM.—HANDBOOK OF CHEMISTRY, Theoretical, Practical, and Technical, with a Recommendatory Preface by Dr. Hofmann. In one large octavo volume of 662 pages, with illustrations.

ASHWELL (SAMUEL), M.D.—A PRACTICAL TREATISE ON THE DISEASES PECULIAR TO WOMEN. Illustrated by Cases derived from Hospital and Private Practice. Third American edition. In one octavo volume of 520 pages.

ARNOTT (NEILL), M.D.—ELEMENTS OF PHYSICS; or, Natural Philosophy, General and Medical. Written for universal use, in plain or non-technical language. A new edition, by Isaac Hays, M.D. Complete in one octavo volume, of 484 pages, with about two hundred illustrations.

BROWN (ISAAC BAKER), M.D.—ON SOME DISEASES OF WOMEN ADMITTING OF SURGICAL TREATMENT. With handsome illustrations. 1 volume, 8vo., extra cloth.

BENNETT (J. HUGHES), M.D.—THE PATHOLOGY AND TREATMENT OF PULMONARY TUBERCULOSIS, and on the Local Medication of Pharyngeal and Laryngeal Diseases, frequently mistaken for, or associated with, Phthisis. In one handsome octavo volume, with beautiful wood-cuts.

BENNETT (HENRY), M.D.—A PRACTICAL TREATISE ON INFLAMMATION OF THE UTERUS, ITS CERVIX AND APPENDAGES, and on its Connection with Uterine Disease. Fourth American, from the third and revised London edition. To which is added (*July*, 1856), A REVIEW OF THE PRESENT STATE OF UTERINE PATHOLOGY. In one neat octavo volume, of 500 pages, with wood-cuts.

BEALE (LIONEL JOHN), M.R.C.S.—THE LAWS OF HEALTH IN RELATION TO MIND AND BODY. A Series of Letters from an old Practitioner to a Patient. In one handsome volume, royal 12mo, extra cloth.

BILLING (ARCHIBALD) M.D.—THE PRINCIPLES OF MEDICINE. Second American, from the fifth and improved London edition. In one handsome octavo volume, extra cloth, 250 pp.

BARCLAY (A. W.), M.D.—A MANUAL OF MEDICAL DIAGNOSIS; being an Analysis of the Signs and Symptoms of Disease. In one neat octavo volume, extra cloth, of over 400 pages. (*Now Ready.*)

BUCKNILL (J. C.), M.D., and DANIEL H. TUKE, M.D.—A MANUAL OF PSYCHOLOGICAL MEDICINE; containing the History, Nosology, Description, Statistics, Diagnosis, Pathology, and Treatment of Insanity. With a Plate. In one handsome octavo volume, of 536 pages. (*Now Ready.*)

BLAKISTON (PEYTON), M. D.—PRACTICAL OBSERVATIONS ON CERTAIN DISEASES OF THE CHEST, and on the Principles of Auscultation. In one volume, 8vo., 384 pages.

BURROWS (GEORGE), M. D.—ON DISORDERS OF THE CEREBRAL CIRCULATION, and on the Connection between the Affections of the Brain and Diseases of the Heart. In one 8vo. vol., with colored plates, pp. 216.

BUDD (GEORGE), M. D.—ON DISEASES OF THE LIVER. Third American, from the third and enlarged London edition. In one very handsome octavo volume, with four beautifully colored plates, and numerous wood-cuts. 500 pages. New edition.

BUDD (GEORGE), M. D.—ON THE ORGANIC DISEASES AND FUNCTIONAL DISORDERS OF THE STOMACH. In one neat octavo volume, extra cloth. (*Now Ready.*)

BUCKLER (T. H.), M. D.—ON THE ETIOLOGY, PATHOLOGY, AND TREATMENT OF FIBRO-BRONCHITIS AND RHEUMATIC PNEUNOMIA. In one handsome octavo volume, extra cloth.

BUSHNAN (J. S.), M. D.—PRINCIPLES OF ANIMAL AND VEGETABLE PHYSIOLOGY. A Popular Treatise on the Functions and Phenomena of Organic Life. In one handsome royal 12mo. volume, extra cloth, with numerous illustrations.

BLOOD AND URINE (MANUALS ON).—BY JOHN WILLIAM GRIFFITH, G. OWEN REESE, AND ALFRED MARKWICK. One thick volume, royal 12mo., extra cloth, with plates. 460 pages.

BRODIE (SIR BENJAMIN C.), M. D.—CLINICAL LECTURES ON SURGERY. One vol., 8vo., cloth. 350 pages.

BIRD (GOLDING), M. D.—URINARY DEPOSITS: THEIR DIAGNOSIS, PATHOLOGY, AND THERAPEUTICAL INDICATIONS. A new and enlarged American, from the last improved London edition. With over sixty illustrations. In one royal 12mo. volume, extra cloth.

BARTLETT (ELISHA), M. D.—THE HISTORY, DIAGNOSIS, AND TREATMENT OF THE FEVERS OF THE UNITED STATES. Fourth edition, revised, with Additions by Alonzo Clark, M. D. In one handsome octavo volume, of 600 pages. (*Just Issued.*)

BOWMAN (JOHN E.), M. D.—PRACTICAL HANDBOOK OF MEDICAL CHEMISTRY. Second American, from the third and revised London edition. In one neat volume, royal 12mo., with numerous illustrations. 288 pages.

BOWMAN (JOHN E.), M. D.—INTRODUCTION TO PRACTICAL CHEMISTRY, INCLUDING ANALYSIS. Second American, from the second and revised English edition. With numerous illustrations. In one neat volume, royal 12mo. 350 pages.

BARLOW (GEORGE H.), M. D.—A MANUAL OF THE PRACTICE OF MEDICINE. With Additions by D. F. Condie, M. D. In one handsome octavo volume, leather, of 600 pages.

CURLING (T. B.), F. R. S.—A PRACTICAL TREATISE ON DISEASES OF THE TESTIS, SPERMATIC CORD, AND SCROTUM. Second American, from the second and enlarged English edition. With numerous illustrations. In one handsome octavo volume, extra cloth.

COLOMBAT DE L'ISERE.—A TREATISE ON THE DISEASES OF FEMALES, and on the Special Hygiene of their Sex. Translated, with many Notes and Additions, by C. D. Meigs, M. D. Second edition, revised and improved. In one large volume, octavo, with numerous woodcuts. 720 pages.

COPLAND (JAMES), M. D.—OF THE CAUSES, NATURE, AND TREATMENT OF PALSY AND APOPLEXY, and of the Forms, Seats, Complications, and Morbid Relations of Paralytic and Apoplectic Diseases. In one volume, royal 12mo., extra cloth. 326 pages

CARSON (JOSEPH), M. D.—SYNOPSIS OF THE COURSE OF LECTURES ON MATERIA MEDICA AND PHARMACY, delivered in the University of Pennsylvania. Second edition, revised. In one very neat octavo volume, of 208 pages.

CARPENTER (WILLIAM B.), M. D.—PRINCIPLES OF HUMAN PHYSIOLOGY; with their chief applications to Psychology, Pathology, Therapeutics, Hygiene, and Forensic Medicine. A new American, from the last and revised London edition. With nearly three hundred illustrations. Edited, with Additions, by Francis Gurney Smith, M. D., Professor of the Institutes of Medicine in the Pennsylvania Medical College, etc. In one very large and beautiful octavo volume, of about 900 large pages, handsomely printed, and strongly bound in leather, with raised bands. *(Just Issued.)*

CARPENTER (WILLIAM B.), M. D.—PRINCIPLES OF COMPARATIVE PHYSIOLOGY. New American, from the fourth and revised London edition. In one large and handsome octavo volume, with over three hundred beautiful illustrations.

CARPENTER (WILLIAM B.), M. D.—THE MICROSCOPE AND ITS REVELATIONS. With an Appendix containing the Applications of the Microscope to Clinical Medicine, by F. G. Smith, M. D. With 434 beautiful wood engravings. In one large and very handsome octavo volume of 724 pages, extra cloth or leather. *(Now Ready.)*

CARPENTER (WILLIAM B.), M. D.—ELEMENTS (OR MANUAL) OF PHYSIOLOGY, INCLUDING PHYSIOLOGICAL ANATOMY. Second American, from a new and revised London edition. With one hundred and ninety illustrations. In one very handsome octavo volume.

CARPENTER (WILLIAM B.), M. D.—PRINCIPLES OF GENERAL PHYSIOLOGY, INCLUDING ORGANIC CHEMISTRY AND HISTOLOGY. With a General Sketch of the Vegetable and Animal Kingdom. In one large and handsome octavo volume, with several hundred illustrations. *(Preparing.)*

CARPENTER (WILLIAM B.), M. D.—A PRIZE ESSAY ON THE USE OF ALCOHOLIC LIQUORS IN HEALTH AND DISEASE. New edition, with a Preface by D. F. Condie, M. D., and explanations of scientific words. In one neat 12mo. volume.

CHRISTISON (ROBERT), M. D.—A DISPENSATORY; or, Commentary on the Pharmacopœias of Great Britain and the United States: comprising the Natural History, Description, Chemistry, Pharmacy, Actions, Uses, and Doses of the Articles of the Materia Medica. Second edition, revised and improved, with a Supplement containing the most important New Remedies. With copious Additions, and two hundred and thirteen large wood-engravings. By R. Eglesfeld Griffith, M. D. In one very large and handsome octavo volume, of over 1000 pages.

CHELIUS (J. M.), M. D.—A SYSTEM OF SURGERY. Translated from the German, and accompanied with additional Notes and References, by John F. South. Complete in three very large octavo volumes, of nearly 2200 pages, strongly bound, with raised bands and double titles.

CONDIE (D. F.), M. D.—A PRACTICAL TREATISE ON THE DISEASES OF CHILDREN. Fifth edition, revised and augmented. In one large volume, 8vo., of over 750 pages. *(Now Ready.)*

COOPER (BRANSBY B.), M. D.—LECTURES ON THE PRINCIPLES AND PRACTICE OF SURGERY. In one very large octavo volume, of 750 pages.

COOPER (SIR ASTLEY P.)—A TREATISE ON DISLOCATIONS AND FRACTURES OF THE JOINTS. Edited by Bransby B. Cooper, F. R. S., etc. With additional Observations by Prof J. C. Warren. A new American edition. In one octavo volume, with numerous wood-cuts.

COOPER (SIR ASTLEY P.)—ON THE STRUCTURE AND DISEASES OF THE TESTIS, AND ON THE THYMUS GLAND. One vol. imperial 8vo., with 177 figures, on 29 plates.

COOPER (SIR ASTLEY P.)—ON THE ANATOMY AND DISEASES OF THE BREAST, with twenty-five Miscellaneous and Surgical Papers. One large volume, imperial 8vo., with 252 figures, on 36 plates

---

CHURCHILL (FLEETWOOD), M. D.—ON THE THEORY AND PRACTICE OF MIDWIFERY. A new American, from the last and improved English edition. Edited, with Notes and Additions, by D. Francis Condie, M. D., author of a "Practical Treatise on the Diseases of Children," etc. With 139 illustrations. In one very handsome octavo volume, 510 pages.

CHURCHILL (FLEETWOOD), M. D.—ON THE DISEASES OF INFANTS AND CHILDREN. Second American edition, revised and enlarged by the author. With Additions by W. V. Keating, M. D. In one large and handsome volume of 700 pages. (*Now Ready.*)

CHURCHILL (FLEETWOOD), M. D.—ESSAYS ON THE PUERPERAL FEVER, AND OTHER DISEASES PECULIAR TO WOMEN. Selected from the writings of British authors previous to the close of the eighteenth century. In one neat octavo volume, of about 450 pages.

CHURCHILL (FLEETWOOD), M. D.—ON THE DISEASES OF WOMEN; including those of Pregnancy and Childbed. A new American edition, revised by the author. With Notes and Additions, by D. Francis Condie, M. D., author of a "Practical Treatise on the Diseases of Children." In one handsome octavo volume, of 768 pages, with wood-cuts. (*Now Ready.*)

DEWEES (W. P.), M. D.—A COMPREHENSIVE SYSTEM OF MIDWIFERY. Illustrated by occasional Cases and many Engravings. Twelfth edition, with the Author's last Improvements and Corrections. In one octavo volume, of 600 pages.

DEWEES (W. P.), M. D.—A TREATISE ON THE PHYSICAL AND MEDICAL TREATMENT OF CHILDREN. Tenth edition. In one volume, octavo, 548 pages.

DEWEES (W. P.), M. D.—A TREATISE ON THE DISEASES OF FEMALES. Tenth edition. In one volume, octavo, 532 pages, with plates.

DRUITT (ROBERT), M. R. C. S.—THE PRINCIPLES AND PRACTICE OF MODERN SURGERY. A new American, from the improved London edition. Edited by F. W. Sargent, M. D., author of "Minor Surgery," &c. Illustrated with one hundred and ninety-three wood-engravings. In one very handsomely-printed octavo volume, of 576 large pages.

DUNGLISON, FORBES, TWEEDIE, AND CONOLLY.—THE CYCLOPÆDIA OF PRACTICAL MEDICINE: comprising Treatises on the Nature and Treatment of Diseases, Materia Medica and Therapeutics, Diseases of Women and Children, Medical Jurisprudence, &c. &c. In four large super-royal octavo volumes, of 3254 double-columned pages, strongly and handsomely bound.

*** This work contains no less than four hundred and eighteen distinct treatises, contributed by sixty-eight distinguished physicians.

DUNGLISON (ROBLEY), M. D. —MEDICAL LEXICON; a Dictionary of Medical Science, containing a concise Explanation of the various Subjects and Terms of Physiology, Pathology, Hygiene, Therapeutics, Pharmacology, Obstetrics, Medical Jurisprudence, &c. With the French and other Synonymes; Notices of Climate and of celebrated Mineral Waters: Formulæ for various Officinal, Empirical, and Dietetic Preparations, &c. Fifteenth edition, revised. In one very thick octavo volume, of about 1000 large double-columned pages, strongly bound in leather, with raised bands. (*Now Ready.*)

DUNGLISON (ROBLEY), M. D.—THE PRACTICE OF MEDICINE. A Treatise on Special Pathology and Therapeutics. Third edition. In two large octavo volumes, of 1500 pages.

DUNGLISON (ROBLEY), M. D.—GENERAL THERAPEUTICS AND MATERIA MEDICA; adapted for a Medical Text-book. Sixth edition, much improved. With one hundred and eighty-seven Illustrations. In two large and handsomely printed octavo volumes, of about 1100 pages. (*Just Issued.*)

DUNGLISON (ROBLEY), M. D.—NEW REMEDIES, WITH FORMULÆ FOR THEIR PREPARATION AND ADMINISTRATION. Seventh Edition, with extensive Additions. In one very large octavo volume, of 770 pages. (*Now Ready.*)

DUNGLISON (ROBLEY), M. D.—HUMAN PHYSIOLOGY. Eighth edition. Thoroughly revised and extensively modified and enlarged, with over 500 illustrations. In two large and handsomely printed octavo volumes, containing about 1500 pages.

DICKSON (S. H.), M. D.—ELEMENTS OF MEDICINE: a Compendious View of Pathology and Therapeutics, or the History and Treatment of Diseases. In one large and handsome octavo volume of 750 pages, leather. (*Just Issued.*)

DAY (GEORGE E.), M. D.—A PRACTICAL TREATISE ON THE DOMESTIC MANAGEMENT AND MORE IMPORTANT DISEASES OF ADVANCED LIFE. With an Appendix on a new and successful mode of treating Lumbago and other forms of Chronic Rheumatism. One volume octavo, 226 pages.

ELLIS (BENJAMIN), M. D.—THE MEDICAL FORMULARY; being a Collection of Prescriptions, derived from the writings and practice of many of the most eminent physicians of America and Europe. Together with the usual Dietetic Preparations and Antidotes for Poisons. To which is added an Appendix on the Endermic use of Medicines, and on the use of Ether and Chloroform. The whole accompanied with a few brief Pharmaceutic and Medical Observations. Tenth edition, revised and much extended, by Robert P. Thomas, M. D., Professor of Materia Medica in the Philadelphia College of Pharmacy. In one neat octavo volume of 296 pages.

ERICHSEN (JOHN).—THE SCIENCE AND ART OF SURGERY; being a Treatise on Surgical Injuries, Diseases, and Operations. With Notes and Additions by the American editor. Illustrated with over 300 engravings on wood. In one large and handsome octavo volume of nearly 900 closely printed pages.

FLINT (AUSTIN), M. D.—PHYSICAL EXPLORATION AND DIAGNOSIS OF DISEASES AFFECTING THE RESPIRATORY ORGANS. In one handsome octavo volume, extra cloth, of 636 pages. (*Now Ready.*)

FERGUSSON (WILLIAM), F. R. S.—A SYSTEM OF PRACTICAL SURGERY. Fourth American, from the third and enlarged London edition. In one large and beautifully printed octavo volume of about 700 pages, with 393 handsome illustrations.

FRICK (CHARLES), M. D.—RENAL AFFECTIONS: their Diagnosis and Pathology. With illustrations. One volume, royal 12mo., extra cloth.

FOWNES (GEORGE), Ph. D.—ELEMENTARY CHEMISTRY, Theoretical and Practical. With numerous Illustrations. Edited, with Additions, by Robert Bridges, M. D. In one large royal 12mo. volume, of over 550 pages, with 181 wood-cuts. Sheep, or extra cloth.

FISKE FUND PRIZE ESSAYS ON CONSUMPTION.—By EDWIN LEE and EDWARD WARREN, M. D In one neat octavo volume, extra cloth. (*Now Ready.*)

GRAHAM (THOMAS), F. R. S.—THE ELEMENTS OF INORGANIC CHEMISTRY. Including the Application of the Science to the Arts. With numerous illustrations. With Notes and Additions, by Robert Bridges, M.D., etc., etc. Second American, from the second and enlarged London edition. 1 vol. 8vo., of over 800 pages, extra cloth. $4.00.

PART I. (*lately issued*), large 8vo., 430 pages, 185 illustrations. $1.50.
PART II. (*now ready*), to match, over 400 pages. $2.50.

GROSS (SAMUEL D.), M. D.—A PRACTICAL TREATISE ON THE DISEASES, INJURIES, AND MALFORMATIONS OF THE URINARY BLADDER, THE PROSTATE GLAND, AND THE URETHRA. Second edition, revised and much enlarged, with 184 illustrations. In one very large and handsome octavo volume, of over 900 pages, extra cloth or leather. (*Just Issued.*)

GROSS (SAMUEL D.), M.D.—A PRACTICAL TREATISE ON FOREIGN BODIES IN THE AIR-PASSAGES. In one handsome octavo volume, with illustrations.

GROSS (SAMUEL D.), M. D. — ELEMENTS OF PATHOLOGICAL ANATOMY; illustrated with three hundred and fifty wood-cuts. Third and revised edition. In one large octavo volume, of over 900 pages, leather.

GROSS (SAMUEL D.), M. D.—A SYSTEM OF SURGERY; Diagnostic, Pathological, Therapeutic, and Operative. With very numerous engravings on wood. (*Preparing.*)

GLUGE (GOTTLIEB), M. D.—AN ATLAS OF PATHOLOGICAL HISTOLOGY. Translated, with Notes and Additions, by Joseph Leidy, M. D., Professor of Anatomy in the University of Pennsylvania. In one volume, very large imperial quarto, with 320 figures, plain and colored, on twelve copper-plates.

GRIFFITH (ROBERT E.), M. D.—A UNIVERSAL FORMULARY, containing the methods of Preparing and Administering Officinal and other Medicines. The whole adapted to Physicians and Pharmaceutists. Second edition, thoroughly revised, with numerous Additions, by Robert P. Thomas, M. D., Professor of Materia Medica in the Philadelphia College of Pharmacy. In one large and handsome octavo volume of over 600 pages, double columns.

GRIFFITH (ROBERT E.), M. D. — MEDICAL BOTANY; or, a Description of all the more important Plants used in Medicine, and of their properties, Uses, and Modes of Administration. In one large octavo volume of 704 pages, handsomely printed, with nearly 350 illustrations on wood.

GARDNER (D. PEREIRA), M. D.—MEDICAL CHEMISTRY, for the use of Students and the Profession: being a Manual of the Science, with its Applications to Toxicology, Physiology, Therapeutics, Hygiene, &c. In one handsome royal 12mo. volume, with illustrations.

HARRISON (JOHN), M. D. — AN ESSAY TOWARDS A CORRECT THEORY OF THE NERVOUS SYSTEM. In one octavo volume, 292 pages.

HUGHES (H. M.), M. D. — A CLINICAL INTRODUCTION TO THE PRACTICE OF AUSCULTATION, and other Modes of Physical Diagnosis, in Diseases of the Lungs and Heart. Second American from the second and improved London edition. In one royal 12mo. volume.

HORNER (WILLIAM E.), M. D. — SPECIAL ANATOMY AND HISTOLOGY. Eighth edition. Extensively revised and modified. In two large octavo volumes, of more than 1000 pages, handsomely printed, with over 300 illustrations.

HOBLYN (RICHARD D.), A. M.—A DICTIONARY OF THE TERMS USED IN MEDICINE AND THE COLLATERAL SCIENCES. Second and improved American edition. Revised, with numerous Additions, from the second London edition, by Isaac Hays, M. D., &c. In one large royal 12mo. volume, of over 500 pages, double columns. (*Now Ready.*)

HABERSHON (S. O.), M. D.—PATHOLOGICAL AND PRACTICAL OBSERVATIONS ON DISEASES OF THE ALIMENTARY CANAL, ŒSOPHAGUS, STOMACH, CÆCUM, AND INTESTINES. With illustrations on wood. In one handsome octavo volume.

*** Publishing in the "Medical News and Library" for 1858.

HAMILTON (FRANK H.)—A TREATISE ON FRACTURES AND DISLOCATIONS. In one handsome octavo volume. With numerous illustrations. (*Preparing.*)

HOLLAND (SIR HENRY), M. D.—MEDICAL NOTES AND REFLECTIONS. From the third London edition. In one handsome octavo volume, extra cloth, of about 500 pages. (*Just Issued.*)

JONES (T. WHARTON), F. R. S.—THE PRINCIPLES AND PRACTICE OF OPHTHALMIC MEDICINE AND SURGERY. Second American, from the second and revised English edition. With Additions by Edward Hartshorne, M. D. In one very neat volume, large royal 12mo., of 500 pages, with 110 illustrations.

JONES (C. HANDFIELD), F. R. S., AND EDWARD H. SIEVEKING, M. D. — A MANUAL OF PATHOLOGICAL ANATOMY. With 397 engravings on wood. In one handsome volume, octavo, of nearly 750 pages, leather. (*Lately Issued.*)

KIRKES (WILLIAM SENHOUSE), M. D., AND JAMES PAGET, F. R. S. — A MANUAL OF PHYSIOLOGY. Third American, from the third and improved London edition. With 200 illustrations. In one large and handsome royal 12mo. volume. 586 pages. (*Now Ready.*)

KNAPP (F.), PH. D.—TECHNOLOGY; or, Chemistry applied to the Arts and to Manufactures. Edited, with numerous Notes and Additions, by Dr. Edmund Ronalds and Dr. Thomas Richardson. First American edition, with Notes and Additions, by Professor Walter R. Johnson. In two handsome octavo volumes, printed and illustrated in the highest style of art, with about 500 wood-engravings.

LEHMANN (G. C.). — PHYSIOLOGICAL CHEMISTRY. Translated from the second edition by George E. Day, M. D. Edited by R. E. Rogers, M. D. With illustrations selected from Funke's Atlas of Physiological Chemistry, and an Appendix of Plates. Complete in two handsome octavo volumes, extra cloth, containing 1200 pages. With nearly 200 illustrations. (*Just Issued.*)

LEHMANN (G. C.). — MANUAL OF CHEMICAL PHYSIOLOGY. Translated from the German, with Notes and Additions, by J. C. Morris, M. D. With an introductory Essay on Vital Force, by Samuel Jackson, M. D. In one handsome octavo volume, extra cloth, of 336 pages. With numerous illustrations.

LA ROCHE (R.), M. D.—PNEUMONIA; its Supposed Connection, Pathological and Etiological, with Autumnal Fevers, including an Inquiry into the Existence and Morbid Agency of Malaria. In one handsome octavo volume, extra cloth, of 500 pages.

LA ROCHE (R.), M. D.—YELLOW FEVER, considered in its Historical, Pathological, Etiological, and Therapeutical Relations. Including a Sketch of the Disease as it has occurred in Philadelphia from 1699 to 1854, with an Examination of the Connections between it and the Fevers known under the same name in other Parts of Temperate, as well as in Tropical Regions. In two large and handsome octavo volumes, of nearly 1500 pages, extra cloth. (*Just Issued.*)

LAWRENCE (W.), F. R. S. — A TREATISE ON DISEASES OF THE EYE. A new edition, edited, with numerous Additions, and 243 illustrations, by Isaac Hays, M. D., Surgeon to Wills' Hospital, etc. In one very large and handsome octavo volume of 950 pages, strongly bound in leather, with raised bands.

LUDLOW (J. L.), M. D. — A MANUAL OF EXAMINATIONS UPON ANATOMY, PHYSIOLOGY, SURGERY, PRACTICE OF MEDICINE, OBSTETRICS, MATERIA MEDICA, CHEMISTRY, PHARMACY, AND THERAPEUTICS. To which is added, a Medical Formulary. Third edition, thoroughly revised and greatly extended and enlarged. With 370 illustrations. In one large and handsome royal 12mo. volume, leather, of over 800 pages. (*Just Issued.*)

LAYCOCK (THOMAS), M. D.—LECTURES ON THE PRINCIPLES AND METHODS OF MEDICAL OBSERVATION AND RESEARCH. In one very neat royal 12mo. volume, extra cloth. (*Just Issued.*)

LALLEMAND (M.).—THE CAUSES, SYMPTOMS, AND TREATMENT OF SPERMATORRHŒA. Translated and edited by Henry J. McDougal. In one volume, octavo, of 320 pages. Second American edition.

LARDNER (DIONYSIUS), D. C. L.—HANDBOOKS OF NATURAL PHILOSOPHY AND ASTRONOMY. Revised, with numerous Additions, by the American editor. The whole complete in three volumes, 12mo., of about 2000 large pages, with over 1000 figures on steel and wood.

MORLAND (W. W.), M. D.—DISEASES OF THE URINARY ORGANS; a Compendium of their Diagnosis, Pathology, and Treatment. With Illustrations. In one large and handsome octavo volume, of about 600 pages, extra cloth. (*Now Ready*, Oct., 1858.)

MEIGS (CHARLES D.), M. D.—WOMAN: HER DISEASES AND THEIR REMEDIES. A Series of Lectures to his Class. Third and improved edition. In one large and beautifully-printed octavo volume.

MEIGS (CHARLES D.), M. D.—OBSTETRICS: THE SCIENCE AND THE ART. Second edition, revised and improved. With 131 illustrations. In one beautifully-printed octavo volume, of 752 large pages.

MEIGS (CHARLES D.), M. D.—A TREATISE ON ACUTE AND CHRONIC DISEASES OF THE NECK OF THE UTERUS. With numerous plates, drawn and colored from nature, in the highest style of art. In one handsome octavo volume, extra cloth.

MEIGS (CHARLES D.), M. D.—ON THE NATURE, SIGNS, AND TREATMENT OF CHILDBED FEVER, in a Series of Letters addressed to the Students of his Class. In one handsome octavo volume, extra cloth, of 365 pages.

MILLER (JAMES), F. R. S. E.—PRINCIPLES OF SURGERY. Fourth American, from the third and revised Edinburgh edition. In one large and very beautiful volume of 700 pages, with 240 exquisite illustrations on wood.

MILLER (JAMES), F. R. S. E.—THE PRACTICE OF SURGERY. Fourth American, from the third Edinburgh edition. Edited, with Additions. Illustrated by 364 engravings on wood. In one large octavo volume of over 700 pages.

MALGAIGNE (J. F.).—OPERATIVE SURGERY, based on Normal and Pathological Anatomy. Translated from the French, by Frederick Brittan, A. B., M. D. With numerous illustrations on wood. In one handsome octavo volume of nearly 600 pages.

MOHR (FRANCIS), PH. D., AND REDWOOD (THEOPHILUS).—PRACTICAL PHARMACY. Comprising the Arrangements, Apparatus, and Manipulations of the Pharmaceutical Shop and Laboratory. Edited, with extensive Additions, by Prof. William Procter, of the Philadelphia College of Pharmacy. In one handsomely-printed octavo volume, of 570 pages, with over 500 engravings on wood.

MACLISE (JOSEPH).—SURGICAL ANATOMY. Forming one volume, very large imperial quarto. With sixty-eight large and splendid Plates, drawn in the best style, and beautifully colored. Containing 190 Figures, many of them the size of life. Together with copious and explanatory letter-press. Strongly and handsomely bound in extra cloth, being one of the cheapest and best executed Surgical works as yet issued in this country.

Copies can be sent by mail, in five parts, done up in stout covers.

MILLER (HENRY), M. D.—PRINCIPLES AND PRACTICE OF OBSTETRICS; including the Treatment of Chronic Inflammation of the Cervix and Body of the Uterus, considered as a frequent Cause of Abortion. With about 100 engravings on wood. In one very handsome octavo volume of over 600 pages. (*Now Ready*.)

MONTGOMERY (W. F.), M. D., &c. — AN EXPOSITION OF THE SIGNS AND SYMPTOMS OF PREGNANCY. With some other Papers on Subjects connected with Midwifery. From the second and enlarged English edition. With two exquisite coloured plates and numerous wood-cuts. In one very handsome octavo volume, extra cloth, of nearly 600 pages.

MAYNE (JOHN), M. D. — A DISPENSATORY AND THERAPEUTICAL REMEMBRANCER. Comprising the entire lists of Materia Medica, with every Practical Formula contained in the three British Pharmacopœias. In one 12mo. volume, extra cloth, of over 300 large pages.

MACKENZIE (W.), M. D. — A PRACTICAL TREATISE ON DISEASES AND INJURIES OF THE EYE. To which is prefixed an Anatomical Introduction, by T. Wharton Jones. From the fourth revised and enlarged London edition. With Notes and Additions by Addinell Hewson, M. D. In one very large and handsome octavo volume, with numerous wood-cuts and plates. 1028 pages, leather, raised bands. (*Just Issued.*)

NEILL (JOHN), M. D., AND FRANCIS GURNEY SMITH, M. D.—AN ANALYTICAL COMPENDIUM OF THE VARIOUS BRANCHES OF MEDICAL SCIENCE; for the Use and Examination of Students. Second edition, revised and improved. In one very large and handsomely printed royal 12mo. volume of over 1000 pages, with 350 illustrations on wood. Strongly bound in leather, with raised bands.

NEILL (JOHN), M. D.—OUTLINES OF THE NERVES. 1 vol. 8vo., with handsome plates. OUTLINES OF THE VEINS AND LYMPHATICS, 1 vol. 8vo., handsome colored plates.

NELIGAN (J. MOORE), M. D. — ATLAS OF CUTANEOUS DISEASES. In one beautiful quarto volume, extra cloth, with splendid colored plates, presenting nearly one hundred elaborate representations of disease. (*Now Ready.*)

NELIGAN (J. MOORE), M. D. —A PRACTICAL TREATISE ON DISEASES OF THE SKIN. In one neat royal 12mo. volume, of 334 pages.

OWEN (PROF. R.) — ON THE DIFFERENT FORMS OF THE SKELETON. One royal 12mo. volume, with numerous illustrations.

PARKER (LANGSTON).—THE MODERN TREATMENT OF SYPHILITIC DISEASES, BOTH PRIMARY AND SECONDARY; comprising the Treatment of Constitutional and Confirmed Syphilis, by a safe and successful method. With numerous Cases, Formulæ, and Clinical Observations. From the third and entirely rewritten London edition. In one neat octavo volume.

PEREIRA (JONATHAN), M. D. — THE ELEMENTS OF MATERIA MEDICA AND THERAPEUTICS. Third American edition, enlarged and improved by the author; including Notices of most of the Medical Substances in use in the civilized world, and forming an Encyclopædia of Materia Medica. Edited, with Additions, by Joseph Carson, M. D., Professor of Materia Medica and Pharmacy in the University of Pennsylvania. In two very large octavo volumes of 2100 pages, on small type, with over 450 illustrations. (*Now Complete.*)

PARRISH (EDWARD).—AN INTRODUCTION TO PRACTICAL PHARMACY. Designed as a Text-book for the Student, and as a Guide for the Physician and Pharmaceutist. With many Formulæ and Prescriptions. In one handsome octavo volume, extra cloth, of 550 pages, with 243 illustrations.

PEASELEE (E. R.), M. D.—HUMAN HISTOLOGY, in its Applications to Physiology and General Pathology. With 434 illustrations. In one handsome octavo volume. (*Now Ready.*)

PIRRIE (WILLIAM), F.R.S.E.—THE PRINCIPLES AND PRACTICE OF SURGERY. Edited by John Neill, M.D., Demonstrator of Anatomy in the University of Pennsylvania, Surgeon to the Pennsylvania Hospital, &c. In one very handsome octavo volume of 780 pages, with 316 illustrations.

RAMSBOTHAM (FRANCIS H.), M. D.—THE PRINCIPLES AND PRACTICE OF OBSTETRIC MEDICINE AND SURGERY, in reference to the Process of Parturition. A new and enlarged edition, thoroughly revised by the author. With Additions by W. V. Keating, M. D. In one large and handsome imperial octavo volume of 650 pages, strongly bound in leather, with raised bands. With sixty-four beautiful plates, and numerous wood-cuts in the text, containing in all nearly 200 large and beautiful figures. (*Just Issued.*)

RIGBY (EDWARD), M. D.—ON THE CONSTITUTIONAL TREATMENT OF FEMALE DISEASES. In one neat royal 12mo. volume, extra cloth, of about 250 pages. (*Just Issued.*)

RICORD (P.), M. D.—ILLUSTRATIONS OF SYPHILITIC DISEASE. Translated from the French, by Thomas F. Betton, M. D. With the addition of a History of Syphilis, and a complete Bibliography and Formulary of Remedies, collated and arranged by Paul B. Goddard, M. D. With fifty large quarto plates, comprising 117 beautifully colored illustrations. In one large and handsome quarto volume.

RICORD (P.), M. D.—A TREATISE ON THE VENEREAL DISEASE. By John Hunter, F. R. S. With copious Additions, by Ph. Ricord, M. D. Edited, with Notes, by Freeman J. Bumstead, M. D. In one handsome octavo volume, with plates.

RICORD (P.), M. D.—LETTERS ON SYPHILIS, addressed to the Chief Editor of the Union Médicale. With an Introduction, by Amédée Latour. Translated by W. P. Lattimore, M. D. In one neat octavo volume.

ROKITANSKY (CARL).—A MANUAL OF PATHOLOGICAL ANATOMY. Translated from the German by W. E. Swaine, Edward Sieveking, M. D., C. H. Moore, and George E. Day, M. D. Complete, four volumes bound in two, extra cloth, of about 1200 pages. (*Just Issued.*)

RIGBY (EDWARD), M. D.—A SYSTEM OF MIDWIFERY. With Notes and Additional Illustrations. Second American edition. One volume octavo, 422 pages.

ROYLE (J. FORBES), M. D.—MATERIA MEDICA AND THERAPEUTICS; including the Preparations of the Pharmacopœias of London, Edinburgh, Dublin, and of the United States. With many new Medicines. Edited by Joseph Carson, M. D., Professor of Materia Medica and Pharmacy in the University of Pennsylvania. With ninety-eight illustrations. In one large octavo volume of about 700 pages.

SKEY (FREDERICK C.), F. R. S.—OPERATIVE SURGERY. In one very handsome octavo volume of over 650 pages, with about 100 wood-cuts.

SHARPEY (WILLIAM), M. D., JONES QUAIN, M. D., AND RICHARD QUAIN, F. R. S., etc.—HUMAN ANATOMY. Revised, with Notes and Additions, by Joseph Leidy, M. D. Complete in two large octavo volumes, of about 1300 pages. Beautifully illustrated with over 500 engravings on wood.

SMITH (HENRY H.), M. D., AND WILLIAM E. HORNER, M. D.—AN ANATOMICAL ATLAS illustrative of the Structure of the Human Body. In one volume, large imperial octavo, with about 650 beautiful figures.

SMITH (HENRY H.), M. D.—MINOR SURGERY; or, Hints on the Every-day Duties of the Surgeon. With 247 illustrations. Third and enlarged edition. In one handsome royal 12mo. volume of 456 pages

SARGENT (F. W.), M. D.—ON BANDAGING AND OTHER OPERATIONS OF MINOR SURGERY. Second edition, enlarged. In one handsome royal 12mo. volume of nearly 400 pages, with 182 illustrations. (*Just Issued.*)

STILLÉ (ALFRED), M. D.—PRINCIPLES OF THERAPEUTICS. In one handsome volume. (*Preparing.*)

SIMON (JOHN). F. R. S.—GENERAL PATHOLOGY, as conducive to the Establishment of Rational Principles for the Prevention and Cure of Disease. A Course of Lectures delivered at St. Thomas's Hospital during the Summer Session of 1850. In one neat octavo volume.

SMITH (W. TYLER), M. D.—ON PARTURITION, AND THE PRINCIPLES AND PRACTICE OF OBSTETRICS. In one large duodecimo volume of 400 pages.

SMITH (W. TYLER), M. D.—THE PATHOLOGY AND TREATMENT OF LEUCORRHŒA. With numerous illustrations. In one very handsome octavo volume, extra cloth, of about 250 pages.

SOLLY (SAMUEL), F. R. S.—THE HUMAN BRAIN; its Structure, Physiology, and Diseases With a Description of the Typical Forms of the Brain in the Animal Kingdom. From the Second and much enlarged London edition. In one octavo volume, with 120 wood-cuts.

SCHŒDLER (FRIEDRICH), PH. D.—THE BOOK OF NATURE; an Elementary Introduction to the Sciences of Physics, Astronomy, Chemistry, Mineralogy, Geology, Botany, Zoology, and Physiology. First American edition, with a Glossary and other Additions and Improvements; from the second English edition. Translated from the sixth German edition, by Henry Medlock, F.C.S., &c. In one thick volume, small octavo, of about 700 pages, with 679 illustrations on wood. Suitable for the higher schools and private students. (*Now Ready.*)

TAYLOR (ALFRED S.), M. D., F. R. S.—MEDICAL JURISPRUDENCE. Fourth American, from the fifth and improved English edition. With Notes and References to American Decisions, by Edward Hartshorne, M. D. In one large octavo volume of 700 pages. (*Now Ready.*)

TAYLOR (ALFRED S.), M. D.—ON POISONS, IN RELATION TO MEDICAL JURISPRUDENCE AND MEDICINE. Edited, with Notes and Additions, by R. E. Griffith, M. D. In one large octavo volume of 688 pages.

TANNER (T. H.), M. D.—A MANUAL OF CLINICAL MEDICINE AND PHYSICAL DIAGNOSIS. To which is added, The Code of Ethics of the American Medical Association. In one neat volume, small 12mo., extra cloth, or flexible. (*Just Issued.*)

TODD (R. B.), M. D.—CLINICAL LECTURES ON CERTAIN DISEASES OF THE URINARY ORGANS, AND ON DROPSIES. In one octavo volume, extra cloth, of about 300 pages. (*Just Issued.*)

TODD (R. B.), M. D., AND WILLIAM BOWMAN, F. R. S.—PHYSIOLOGICAL ANATOMY AND PHYSIOLOGY OF MAN. Now complete, in one very large and handsome octavo volume, of 926 pages, with 300 illustrations on wood. (*Just Issued*, 1857.)

☞ Gentlemen who have the earlier portions of this work can still complete their copies, if early application be made.

WATSON (THOMAS), M. D., &c.—LECTURES ON THE PRINCIPLES AND PRACTICE OF PHYSIC. A new American, from the last London edition. Revised, with Additions, by D. Francis Condie, M. D., author of a "Treatise on the Diseases of Children," &c. In one imperial octavo volume, of over 1200 large pages, with nearly 200 cuts, strongly bound, with raised bands. (*Now Ready.*)

WALSHE (W. H.), M. D.—DISEASES OF THE HEART, LUNGS, AND APPENDAGES; their Symptoms and Treatment. In one handsome volume, large royal 12mo., 512 pages.

WHAT TO OBSERVE AT THE BEDSIDE AND AFTER DEATH, IN MEDICAL CASES. Published under the authority of the London Society for Medical Observation. In one very handsome volume, royal 12mo., extra cloth.

WILDE (W. R.).—AURAL SURGERY, AND THE NATURE AND TREATMENT OF DISEASES OF THE EAR. In one handsome octavo volume, with illustrations.

WHITEHEAD (JAMES), F. R. C. S., &c.—THE CAUSES AND TREATMENT OF ABORTION AND STERILITY; being the Result of an Extended Practical Inquiry into the Physiological and Morbid Conditions of the Uterus. Second American Edition. In one volume, octavo, 368 pages

---

WEST (CHARLES), M. D.—LECTURES ON THE DISEASES OF INFANCY AND CHILDHOOD. Second American, from the second and enlarged London edition. In one volume, octavo, of nearly 500 pages.

WEST (CHARLES), M. D.—AN INQUIRY INTO THE PATHOLOGICAL IMPORTANCE OF ULCERATION OF THE OS UTERI. Being the Croonian Lectures for the year 1854. In one neat octavo volume, extra cloth.

WEST (CHARLES), M. D. — LECTURES ON THE DISEASES OF WOMEN. In two Parts. Part I., Diseases of the Uterus. Part II., Diseases of the Ovaries, &c., the Bladder, Vagina, and External Organs. Complete in one octavo volume of 500 pages, extra cloth. (*Now Ready.*)
Part II. now ready, 1 vol., 8vo., extra cloth, of about 200 pages. Sold separate, $1.

WILSON (ERASMUS), M. D., F. R. S.—A SYSTEM OF HUMAN ANATOMY, General and Special. Fourth American, from the last English edition. Edited by W. H. Gobrecht, M. D. With 400 illustrations. Beautifully printed, in one large octavo volume, of over 600 pages. (*Now Ready.*)

WILSON (ERASMUS), M. D., F. R. S.—THE DISSECTOR'S MANUAL; Practical and Surgical Anatomy. Third American, from the last revised and enlarged English edition. Modified and re-arranged by William Hunt, M. D. In one large and handsome royal 12mo. volume, leather, of 582 pages, with 154 illustrations.

WILSON (ERASMUS), M. D., F. R. S.—ON DISEASES OF THE SKIN. Fourth American, from the Fourth London edition. In one neat octavo volume, of 650 pages, extra cloth.
Also, AN ATLAS OF PLATES, of which twelve are exquisitely coloured, illustrating "WILSON ON THE SKIN." 8vo., cloth. (*Now Ready.*)

WILSON (ERASMUS), M. D., F. R. S.—ON CONSTITUTIONAL AND HEREDITARY SYPHILUS, AND ON SYPHILITIC ERUPTIONS. In one small octavo volume, beautifully printed, with four exquisite coloured plates, presenting more than thirty varieties of Syphilitic Eruptions.

WILSON (ERASMUS), M. D., F. R. S.—HEALTHY SKIN; a Treatise on the Management of the Skin and Hair in Relation to Health. Second American, from the fourth and improved London edition. In one handsome royal 12mo. volume, extra cloth, with numerous illustrations. Copies may also be had in paper covers, for mailing, price 75 cents.

WILLIAMS (C. J. B.), M.D., F.R.S.—PRINCIPLES OF MEDICINE; comprising General Pathology and Therapeutics, and a brief General View of Etiology, Nosology, Semeiology, Diagnosis, Prognosis, and Hygienics. Fifth American, from a new and enlarged London edition. In one octavo volume, of 500 pages.

YOUATT (WILLIAM), V. S. — THE HORSE. A new edition, with numerous illustrations; together with a General History of the Horse; a Dissertation on the American Trotting Horse; how Trained and Jockeyed; an account of his Remarkable Performances; and an Essay on the Ass and the Mule. By J. S. Skinner, formerly Assistant Postmaster-General, and Editor of the Turf Register. One large octavo volume.

YOUATT (WILLIAM), V. S.—THE DOG. Edited by E. J. Lewis, M. D. With numerous and beautiful illustrations. In one very handsome volume, crown 8vo., crimson cloth, gilt.

Illustrated Catalogue.

Blanchard & Lea have now ready a detailed Catalogue of their publications, in Medical and other Sciences, with Specimens of the Wood-engravings, Notices of the Press, &c. &c., forming a pamphlet of eighty large octavo pages. It has been prepared without regard to expense, and may be considered as one of the handsomest specimens of printing as yet executed in this country. Copies will be sent free, by post, on receipt of nine cents in postage stamps.

Detailed Catalogues of their publications, Miscellaneous, Educational, Medical, &c., furnished gratis, on application.

www.ingramcontent.com/pod-product-compliance
Lightning Source LLC
LaVergne TN
LVHW010205110826
845151LV00002B/610
* 9 7 8 1 4 2 5 5 3 6 0 5 3 *